Potassium

A & W Basics in Medicine Series

Potassium

Horacio J. Adrogué, M.D.

Professor of Medicine
Baylor College of Medicine
Chief, Renal Section
Veterans Affairs Medical Center
Houston, Texas

Donald E. Wesson, M.D.

Associate Professor of Medicine
Baylor College of Medicine
Assistant Chief, Renal Section
Veterans Affairs Medical Center
Houston, Texas

Libra & Gemini Publications, Inc.
Houston, Texas

A & W Basics in Medicine Series. Potassium
Copyright © 1992 by Libra & Gemini Publications, Inc.
First Edition

All rights reserved. No part of this book may be reproduced in any form or by any electronic or mechanical means including information storage and retrieval systems without permission in writing from the publisher or authors, except by a reviewer who may quote brief passages in a review.

Notice of Liability

The authors and publisher have used their best efforts in preparing this book. While every precaution has been taken, neither the authors nor the publisher shall have any liability to any person or entity with respect to damage caused or alleged to be caused directly or indirectly by the instructions and suggestions contained in this book.

Library of Congress Catalog Card Number 92-90173

ISBN 0-9630670-1-X

ISSN 1057-9575

Printed in the United States of America
Libra & Gemini Publications, Inc.
10703 Paulwood
Houston, Texas 77071
713/981-6321

Dedication

To our families

Sarita
Sofía, Horacio, Soledad,
Matías and Marcos Adrogué

Wanda
David and Donald Wesson

Table of Contents

Preface   *xi*

Issues
    Published and Forthcoming   *xiii*

Acknowledgements   *xv*

Basic Concepts   *3*

Mechanisms of K^+ Homeostasis   *47*

Clinical Disorders
    1. Hypokalemia   *89*
    2. Hyperkalemia   *167*

Selected References   *223*

Index   *225*

Preface

The A & W Basics in Medicine Series examines relevant topics in Medicine using concepts that pertain to the basic sciences, as a starting point. The doctrine on which this series has been formulated dictates that two critical elements should be present in order to acquire the skills required for optimal patient management. These two elements are: a thoughtful approach based on solid pathophysiological knowledge and the ability to apply common sense generously while acquiring basic medical principles. If properly combined, the presence of these two elements allows the unraveling of difficult and/or obscure clinical situations. This series uses simple questions followed by short answers as well as case presentations to illustrate the various topics presented. The sequential order of the questions chosen provides the reader with layers of knowledge that will steadily increase his/her understanding of each topic. The application of the various concepts mentioned above in the presentation of each case will help to consolidate theoretical principles with practical facts.

Each issue of a book of this series examines a carefully selected topic that is germane to the daily practice of medicine. In it, the readers will find the tools for an intelligent interpretation of clinical data based on sound pathophysiological concepts enabling them to provide quality care.

The teaching approach used in this series is at variance with that of the traditional system of medical education that advocates the acquisition of an encyclopedic knowledge as the critical weapon to deal with medical problems. Our belief is, instead, that the time and energy spent by science students in knowledge acquisition should be limited and the emphasis should be placed in expanding the student's ability to integrate theory and practice.

H.J.A.
D.E.W.

Issues

Published

- ✧ Acid-Base
- ✧ Potassium

Forthcoming

- ✧ Salt & Water
- ✧ Heart Failure
- ✧ Renal Failure

Acknowledgements

The authors would like to acknowledge the excellent secretarial assistance of Debby S. Verrett and the expert advice and assistance of María T. Coggins and Lucrecia Hug in the various phases necessary for the completion of this book.

Basic Concepts

1 What is the role of potassium ions in biological systems?

❏ Potassium ions (K^+) have a critical role in many vital cell functions such as its metabolism, growth, repair, and volume regulation as well as in the electric properties of the cell.

2 Is the key role of K^+ limited to the cell biology of higher animals?

❏ No. Potassium ions are essential constituents in all forms of life including bacteria, plants, and animals.

3 Compare the ionic composition of the ancient sea and that of intracellular fluid (ICF) of current forms of life.

❏ The high concentration of K^+ that is present in ICF was also a characteristic feature of the ionic composition of the ancient sea.

4 Describe the possible linkage between the composition of the original sea and that of intracellular fluid (ICF).

❏ Primitive life forms might have corralled a portion of sea water that became its interior environment or "milieu interne" (as described by Claude Bernard).

5 What importance might have had a K^+-rich internal environment in the evolution of life forms?

❏ Potassium ions stimulate many cellular enzymes and the K^+-rich cell interior provides an optimal ionic composition for the enzymatic machinery of cells. This K^+-rich cell interior was important as simple life forms evolved to more complex ones. Therefore, cells have developed effective mechanisms that sustain a high intracellular concentration of potassium ($[K^+]_i$) in comparison with that of the surrounding medium.

6 What influence might have had the K^+ accumulation in living organisms upon the electrolyte composition of the ancient sea?

❏ The incorporation of K^+ into the cells of organisms living in the sea changed the sea composition from its original high K^+ concentration to its current high sodium/low K^+ levels.

7 Which is the most prevalent cation in the intracellular fluid (ICF)?

❏ The most abundant cation in the cell is potassium. Its concentration in ICF is ~150 meq/liter. This value is typically found in muscle tissue.

8 Which are the other major cations in the intracellular fluid?

❑ The other major intracellular cations are magnesium (Mg^{++}) and sodium (Na^+). The concentration of these ions is approximately 40 meq/liter and 10 meq/liter, respectively.

9 What is the concentration of Na^+ and K^+ in extracellular fluid (ECF)?

❑ The extracellular concentration of sodium ($[Na^+]_e$) and potassium ($[K^+]_e$) is ~140 meq/liter and 5 meq/liter, respectively.

10 Which are the other major cations in the extracellular fluid?

❑ The other major extracellular cations are calcium (Ca^{++}) and (Mg^{++}). The concentration of these ions is ~5 meq/liter and 2 meq/liter, respectively.

11 Give the proportion of K^+ present in intracellular fluid, compared to that in extracellular fluid (ICF and ECF).

❑ The bulk of K^+ is located within the cells and amounts to 98% of the total body K^+. Only 2% of the K^+ is present in ECF.

12 What is the total K^+ content in humans, expressed in meq/kg body wt?

❏ Direct chemical analysis reveals K^+ levels of ~55 meq/kg body wt in adults. Indirect methods, including total-body counting of $^{40}K^+$ and the dilution technique using $^{42}K^+$, yield lower values, ~50 meq/kg body wt.

13 Describe the K^+ distribution in the body tissues of a 70 kg person.

❏ The total body K^+ amounts to 3,500 meq (using 50 meq/kg body wt) and by far the largest fraction (~77%) is held within the skeletal muscle (~2,700 meq). The K^+ content of the liver, bone, and red cells is, in each of these tissues, ~10% (~270 meq) of the amount found in the skeletal muscle.

14 How much K^+ is held in ECF?

❏ ECF contains ~70 meq of potassium, which represents 2% of the total body K^+ content (3,500 meq).

15 What determines the unequal distribution of K^+ between the ICF and ECF?

❏ The unequal distribution of K^+ is due to the additive effects of the Gibbs-Donnan rule and the "pump-leak" hypothesis.

16 Describe the Gibbs-Donnan rule applied to the distribution of ions between the ICF and ECF.

❏ Large macromolecules having a net negative charge are confined to the interior of the cell membrane (intracellular compartment). These cell-restricted anions are responsible for the peculiar distribution of the diffusible cations and anions on each side of the cell membrane. The Gibbs-Donnan rule allows precise quantification of the uneven distribution of ions in ICF and ECF.

17 What are these cell-restricted large macromolecules?

❏ These polyvalent macromolecules are phosphate compounds and proteins, essential in normal cell function, abundantly present in the cytosolic fluid contained in all cells. At a normal cell pH, these macromolecules have a net negative charge that is responsible for more than two-thirds of the anionic equivalency of ICF.

18 What is the concentration of the major intracellular anions in the cytosolic fluid?

❏ Phosphates ($HPO_4^=$) and proteins ($Prot^-$) comprise approximately 100 meq/liter and 55 meq/liter of the intracellular anionic equivalency, respectively. Their additive value accounts for more than 90% of the total intracellular anionic equivalency.

19 What are the diffusible ions whose distribution is modulated by the cell-restricted anions?

❑ In most cells, K^+ (cation) and chloride (Cl^-) (anion) are the dominant conductive species (diffusible ions whose movement across the membrane carries a net electric charge across the cell membrane) and, in accordance with the Gibbs-Donnan rule, these permeant ions must be in electrochemical equilibrium.

20 What is the dominant intracellular cation that balances the negative charge of the cell-restricted macromolecules?

❑ The dominant intracellular cation is potassium. Since K^+ is the exclusive cation with substantial permeability across the cell membrane, it will accumulate within the cytosol, balancing the negative charge of $HPO_4^=$ and $Prot^-$.

21 What is the nature of the interaction among K^+, $HPO_4^=$ and $Prot^-$, in the ICF?

❑ Intracellular $HPO_4^=$ and $Prot^-$ interact with K^+ almost exclusively by electrostatic binding.

22 What is the major cation in ECF?

❑ The major cation in ECF is Na^+, as opposed to K^+, which is the dominant cation in ICF.

23 Is Na^+ an impermeant ion species across cell membranes?

❏ No. Although Na^+ is not strictly impermeant, its rapid extrusion upon entering the cell makes this cation effectively impermeant.

24 What is the effect of the negatively charged intracellular macromolecules on the distribution of Cl^-?

❏ The cell-restricted anions force an unequal distribution of Cl^-, the permeant anion, which largely accumulates in ECF. It is apparent that a cation must balance the negative charge of Cl^- outside the cell; the cation that fulfills this function is Na^+.

25 What is the mathematical relationship, if any, between the concentration of diffusible cations and anions in ICF and ECF?

❏ The Gibbs-Donnan rule dictates that the product of the concentration of the diffusible cations and anions on each side of the membrane is identical. Thus, the product of the concentration of the diffusible ions confined to ICF is identical to the product of the concentration of the diffusible ions found in ECF.

26 Apply the above-mentioned mathematical relationship to explain the distribution of K^+ and Cl^- across cell membranes in qualitative terms.

❏ The Gibbs-Donnan rule establishes that ICF must have a high concentration of K^+ and a low concentration of Cl^-. Conversely, ECF must have a low concentration of K^+ and a high concentration of Cl^-. The peculiar distribution of these two ionic species is due to the high and exclusive permeability of the cell membrane to K^+ and Cl^-.

27 Apply the above-mentioned mathematical relationship to explain the quantitative distribution of K^+ and Cl^- across cell membranes.

❏ This relationship will be examined in cytosolic fluid from skeletal muscle and ECF. Since the concentration of K^+ and Cl^- in skeletal muscle cells (ICF) is 150 meq/liter and 4 meq/liter, respectively, the product of these diffusible ions is 600. In a comparable manner, we multiply the concentration of Cl^- and K^+ in ECF, with levels of 120 meq/liter and 5 meq/liter, respectively; the product of the concentration of these diffusible ions in ECF is 600, a value identical to the one obtained in ICF.

28 Explain the reasons why the Gibbs-Donnan rule is not the exclusive mechanism responsible for the unique distribution of K^+ and Na^+ across the cell membrane.

❏ The largely extracellular distribution of Na^+ is not simply the result of a nonpermeant condition of this ion across the cell membrane but is also explained by active pumping of Na⁺ from cells by the Na⁺, K⁺-ATPase (Na⁺, K⁺-pump). Inhibition of this pumping mechanism by ouabain, a classic digitalis preparation, results in Na⁺ accumulation and K⁺ loss from the cell.

29 Give additional evidence to prove that the Gibbs-Donnan rule is not the exclusive mechanism responsible for the particular distribution of K⁺ and Na⁺ across the cell membrane.

❑ Precise measurements of the concentration of intracellular K⁺ ($[K^+]_i$) with ion-selective microelectrodes in many cell types reveal a higher $[K^+]_i$ than the one predicted by the Gibbs-Donnan rule. The higher than expected $[K^+]_i$ is partially explained by active transport mechanisms (e.g., Na⁺, K⁺-ATPase) across the cell membrane, that require cell energy expenditure.

30 Describe the basic elements of the "pump-leak" hypothesis.

❑ The "pump-leak" hypothesis establishes that a steady-state intracellular concentration of a given ion (i.e., K⁺) develops when the rates of the two opposing processes, pump and leak, are equal. The pump generates a high concentration by translocating ions from one compartment to the other. The leak dissipates the pump-created gradient by allowing the movement of ions in the opposite direction. While pumps move ions against their physicochemical gradient, leaks allow equilibration of ions according to passive electrochemical gradients.

31 What is the effect of the cellular accumulation of K^+ and exclusion of Cl^- on the electric properties of the cell?

❏ The high $[K^+]_i$ and low $[Cl^-]_i$ generate an electric potential difference (PD) across the membrane where the cytosol is negative with respect to the ECF. The PD counterbalances the chemical gradient across the cell membrane and therefore satisfies the electrochemical equilibrium of the permeant ions, K^+ and Cl^-.

32 Describe the relationship between chemical gradient and the resulting PD using the Nernst equation.

❏ Nernst established that a PD will develop if a membrane with a selective permeability to either the cation (i.e., K^+) or the anion (i.e., $HPO_4^=$) separates two solutions with different ionic concentrations. If a membrane allows the movement of K^+ but not the movement of $HPO_4^=$ (the balancing anion), and if the concentration of K_2HPO_4 in the two solutions differs, a PD will develop between the two compartments. Potassium ions will cross the membrane from the high concentration solution to the low one, thereby creating a negative voltage in the compartment from which they exit.

33 Describe the expected PD resulting from a tenfold concentration difference between the solutions in two compartments separated by a membrane with selective permeability to either the cation or the anion in the cation/anion pair.

❏ The PD generated by the tenfold concentration difference between the solutions is approximately 58 mV.

34 Describe the sign or polarity and the magnitude of the PD generated by a tenfold concentration difference between two solutions of K_2HPO_4 separated by a membrane with selective permeability to K^+.

❏ The PD generated as a result of this concentration difference (i.e., $[K^+]$ of 100 meq/liter in compartment A and 10 meq/liter in compartment B) will be ~58 mV, and the compartment with low $[K^+]$ will become electrically positive with respect to the compartment with high $[K^+]$. Thus, compartment A will be 58 mV electronegative in comparison with compartment B.

35 How was it determined that a tenfold concentration gradient under the conditions previously described rendered a PD of 58 mV?

❏ Investigators measured the PD that resulted from changes in the chemical gradient across an ion-selective membrane and thereby established the relationship between chemical and electric potential.

36 Compare possible differences in the electric profile of a membrane separating solutions with a tenfold concentration gradient when the membrane is: a) ion-selective (for either the cation or the anion); and b) nonselective, that is, a membrane with identical permeability to both the cation and the anion.

❑ The PD that results from the ion selective membrane is 58 mV, as described previously. A PD will not develop when a nonselective membrane separates the two compartments since the chemical gradient between the two compartments is satisfied by the simultaneous movement of particles having opposite electric charges.

37 Define equilibrium potential.

❑ The equilibrium potential is the electric force or PD required to counterbalance the chemical driving force.

38 Give an example of the relationship between chemical driving force and equilibrium potential.

❑ Let us assume two compartments that are separated by a cation-selective permeable membrane (i.e., one permeable to K^+). One compartment has a concentration of potassium chloride (KCl) that is severalfold higher than that in the other compartment. The chemical gradient between the solutions will move K^+ from the compartment of higher concentration to the one of lower concentration. The exit of K^+, a positively charged ion, will generate a PD by reducing the amount of positive charge in the compartment it exits and increasing it in the one it enters. The compartment from which K⁺ exits will become electrically negative compared to the one it enters. This PD, or membrane potential (MP) will counterbalance the tendency of K⁺ to diffuse from the high concentration to the low concentration compartment.

39 Describe the Nernst equation.

❑ The Nernst equation allows quantification of the equilibrium potential resulting from the selective permeability of the membrane to a given ion. The equation is as follows:

$$E_{ion} = \frac{RT}{ZF} \ln \frac{[ion]_e}{[ion]_i}$$

where E_{ion} is the equilibrium potential of the given ion, R is the gas constant, T is the absolute temperature, F is the Faraday constant, ln is the natural logarithm, $[ion]_i$ is the ion concentration in one compartment (e.g., intracellular fluid or ICF), $[ion]_e$ is the ion concentration in the other compartment (e.g., extracellular fluid or ECF), and Z is the charge of the diffusing ion.

40 Describe a "user friendly" form of the Nernst equation.

❑ The first term of the Nernst equation represents a ratio of various constants that can be replaced by the number 25 in the case of monovalent ions. Consequently, the equation becomes:

$$E_{ion} = 25 \ln \frac{[ion]_e}{[ion]_i}$$

41 Describe a "user friendly" form of the Nernst equation that uses a decimal logarithm instead of the natural logarithm used in the above equation.

❏ The utilization of decimal logarithms to express the value of the ratio of the ion concentration in the two compartments mandates the use of a value for the first term of the equation different from 25, the value applied when natural logarithms are used in the equation. Therefore, the equation becomes:

$$E_{ion} = 58 \log \frac{[ion]_e}{[ion]_i}$$

42 Apply the Nernst equation to calculate E_{ion} for the distribution of K^+ in frog skeletal muscle having a $[K^+]_i$ and a $[K^+]_e$ of 139 meq/liter and of 2.5 meq/liter, respectively.

❏ Let us insert the respective values into our "user friendly" form of the Nernst equation, as follows:

111 List the factors that modify internal K⁺ distribution.

❑ The major factors that alter internal K⁺ balance include hormones (insulin, catecholamines), the acidity of body fluids, the levels of other electrolytes, the tonicity of body fluids, and drugs.

112 Describe the overall effects of insulin on plasma K⁺ levels ($[K^+]_p$).

❑ Insulin is a major modulator of extrarenal K⁺ homeostasis and promotes K⁺ uptake in many cell types, including those from skeletal muscle and liver. The hypokalemic action occurs at very low concentrations of insulin and it is independent of the effect of insulin on glucose uptake. The precise mechanism of this action remains to be defined but appears to involve the activation of several transport proteins.

113 Name the various cellular mechanisms of insulin-mediated K⁺ loading.

❑ The cellular mechanisms of insulin-mediated K⁺ loading are: (1) stimulation of the Na⁺, K⁺-ATPase; (2) stimulation of the Na⁺/H⁺ exchanger; and (3) depression of ionic conductance of some K⁺ channels.

114 Describe the hypokalemic effect of insulin resulting from stimulation of the Na^+, K^+-ATPase.

❏ Direct stimulation of the Na^+, K^+-pump by insulin induces the translocation of K^+ to the cell interior (entry of two K^+ and exit of three Na^+). In addition, substantial changes occur in the transport of electrolytes across the cell. The immediate result of insulin-mediated stimulation of Na^+, K^+-ATPase on the electric properties of the cell membrane is hyperpolarization of the MP (a more negative cell interior). The secondary effects promoted by insulin-induced hyperpolarization of the cell membrane are: (1) a new electric gradient, which favors cellular K^+ entry; and (2) deactivation of K^+ channels, which inhibits cellular K^+ exit. Thus, the secondary effects of insulin on the MP increase the hypokalemic action of this hormone.

115 Describe the hypokalemic effect of insulin resulting from stimulation of the Na^+-H^+ exchanger in the cell membrane.

❏ Insulin promotes the cellular entry of Na^+ and the cellular exit of H^+ by stimulation of this exchanger. The entry of Na^+ increases the $[Na^+]_i$ which further stimulates the Na^+, K^+-ATPase. The cellular exit of H^+ results in cytosolic alkalinization which, in turn: (1) increases the K^+-binding capacity of intracellular anions; and (2) stimulates the Na^+, K^+-pump, therefore favoring cellular K^+ loading.

116 Describe the hypokalemic effect of insulin resulting from its action on K⁺ channels.

❑ Insulin controls gating of the inward rectifier K⁺ channel (K⁺Ch) of skeletal muscle. This type of K⁺Ch is of significant relevance since it is responsible for most of the K⁺ conductance of the skeletal muscle in the resting state. This channel allows K⁺ to flow into cells much more easily than to exit cells. Hence, when the cell membrane is hyperpolarized, the high inward conductance facilitates cellular K⁺ entry. When the cell membrane is depolarized, the low outward conductance reduces K⁺ exit from cells. Insulin exaggerates the inward rectifying properties of this class of K⁺Ch by a dual effect of stimulation of K⁺ entry and depression of K⁺ exit.

117 What is the effect of glucagon on $[K^+]_p$?

❑ Glucagon induces glycogen breakdown in the hepatocytes, releasing glucose and K⁺. Consequently, high glucagon levels can elicit a transient increase in $[K^+]_p$. An increase in plasma glucagon levels has been described in acute metabolic acidosis and this hormonal response might play a role in acidosis-induced hyperkalemia.

118 What is the overall effect of the sympathetic nervous system on internal K⁺ exchanges?

❑ The sympathetic system has a major effect on internal K⁺ homeostasis. Activation of this system might cause a substantial increase or decrease in

$[K^+]_e$. Whether the effect is hypokalemia or hyperkalemia is dependent upon the composite actions mediated by alpha and beta adrenoreceptors.

---

119 Describe the effect of alpha-adrenergic stimulation on $[K^+]_p$.

❏ Alpha-adrenergic stimulation produces hyperkalemia which has been largely attributed to K^+ release from splanchnic tissues, mostly from the liver. During vigorous exercise, an acute major rise in $[K^+]_p$ occurs which is partially caused by alpha-adrenergic stimulation leading to skeletal muscle release of K^+.

---

120 Describe the effect of beta-adrenergic stimulation on $[K^+]_p$.

❏ Beta-adrenergic stimulation causes hypokalemia due to K^+ uptake in skeletal muscle and splanchnic bed. Beta-2 adrenoreceptors are involved in this hypokalemic response that occurs with the endogenous release of catecholamines and with the exogenous administration of drugs with these properties. The role of beta-2 receptors is most evident when considering that epinephrine, a beta-2 agonist, produces substantial hypokalemia; dobutamine on the other hand, a beta-1 selective agonist, does not have a hypokalemic effect.

121 Name two conditions in which the sympathetic nervous system plays a major role in the physiologic regulation of $[K^+]_p$.

❏ The internal disposal of dietary K^+ and the modulation of $[K^+]_p$ during vigorous exercise exemplify conditions in which the sympathetic nervous system plays a major role in the modulation of internal K^+ balance.

122 Describe the relationship between the activity of the sympathetic nervous system, the feeding pattern, and the internal K^+ disposal.

❏ Feeding stimulates the sympathetic nervous system, while fasting inhibits it. A large fraction of dietary K^+ intake is rapidly incorporated into cells by a process stimulated by beta-adrenergic-mediated K^+ disposal, thereby preventing a major postprandial rise in $[K^+]_p$. It should be noted that the dietary-induced release of insulin is also involved in preventing a major rise in $[K^+]_p$ that may be otherwise observed when a large dietary K^+ intake occurs.

123 Describe the effects of the sympathetic nervous system on $[K^+]_p$ during vigorous exercise.

❏ Vigorous exercise causes sympathetic stimulation and release of catecholamines into the circulation. In addition, exercise releases K^+ from the working muscle to ECF. This release of K^+ during exercise, as well as K^+ uptake during the recovery post-exercise, is modulated by the sympathetic nervous system. The role of the sympathetic nervous system is evident during exercise, with simultaneous intake of adrenergic blockers; beta blockade (propranolol) increases $[K^+]_p$ and alpha blockade (phentolamine) decreases $[K^+]_p$ during exercise. The effects of these adrenergic blockers on $[K^+]_p$ are also evident in the recovery from exercise in which a similar pattern is observed.

124 Describe the effects of changes in the acidity of body fluids on $[K^+]_p$.

❏ Acidemia is generally associated with K^+ release from ICF, increasing $[K^+]_p$. Conversely, alkalemia generally shifts K^+ into cells, thereby decreasing $[K^+]_p$.

125 Does plasma bicarbonate concentration ($[HCO_3^-]_p$) modulate $[K^+]_p$ independently of its effects on extracellular pH?

❏ Yes. Classic studies suggested that the changes in $[K^+]_p$ observed during acute acid-base disorders emanated only from the attendant changes in plasma pH and were independent of the metabolic or respiratory nature of the disorder. More recent studies, however, have shown that acute variations in $[HCO_3^-]_p$ under isohydric conditions (by appropriately manipulating P_aCO_2 the pH is maintained constant) are attended by reciprocal changes in $[K^+]_p$. Thus, elevated $[HCO_3^-]_p$ tends to produce hypokalemia while low $[HCO_3^-]_p$ tends to produce hyperkalemia. It has been proposed that changes in the HCO_3^- gradient between ECF and ICF can produce a transfer of HCO_3^- between these compartments, with K^+ as the accompanying cation.

126 What are the critical determinants of the alteration in $[K^+]_p$ observed in response to changes in blood pH (K^+-H^+ relationship)?

❑ The K^+-H^+ relationship is explained by "basic determinants" as well as by "modulators". The basic determinants are the changes in the anionic binding sites for K^+ in the cytosol, which in turn alter the electrochemical equilibrium for this ion, favoring either the cellular exit of K^+ or its uptake. The modulators of the K^+-H^+ relationship include the action of insulin, the sympathetic nervous system, and the effects of pH on cell membrane K^+ pathways. The importance of the modulators is modest in comparison to that of the basic determinants with respect to the $\Delta[K^+]_p / \Delta pH$ in the various acid-base disorders.

127 Describe a major determinant of K^+ release from cells during acidemic states.

❑ Intracellular proteins interact with K^+ by electrostatic binding and the approximately 55 meq/liter of anionic proteins within skeletal muscle cells are balanced by K^+. The ion-binding sites of proteins have preference for H^+ over other ions; however other ions can compete only if their concentration is considerably higher than that of H^+. This requisite is amply satisfied in the cytosolic fluid by K^+ because its concentration is six orders of magnitude higher than that of H^+ ($[K^+]_i = 10^{-1}$ M;

$[H^+]_i = 10^{-7}$ M). Nevertheless, when cell pH decreases ($[H^+]_i$ increases) hydrogen ions (H^+) displace K^+ from the proteins which are known to have substantial buffer value. The diminished binding of K^+ by intracellular proteins in acidosis will alter the electrochemical gradient for this ion favoring the cellular exit of K^+.

---

128 Explain what significance do changes in the intracellular concentration of bicarbonate during acidemic states have in the release of K^+ from cells.

❑ The intracellular level of HCO_3^- ($[HCO_3^-]_i$) of ~12 meq/l in skeletal muscle balances an equal concentration of cations, most of which are K^+. Thus, increases and decreases in $[HCO_3^-]_i$ should be accompanied by parallel alterations in the levels of K^+ bound to this anion within the cell.

In metabolic acidosis, the decreased $[HCO_3^-]_i$ will diminish the anionic sites for K^+ binding, thereby favoring the cellular exit of K^+.

---

129 Explain what significance do changes in the intracellular concentration of bicarbonate ($[HCO_3^-]_i$) have in the release of K^+ from cells during respiratory acidosis.

❑ In respiratory acidosis intracellular $[HCO_3^-]$ increases, contrasting with its decrease in metabolic acidosis. In addition, K^+ is released from intracellular proteins due to the buffering of hydrogen ions (H^+) in respiratory acidosis; the displaced K^+ can be electrostatically balanced by the anionic equivalency resulting from the simultaneous generation of HCO_3^-. Consequently, in respiratory acidosis there is not a substantial release of cellular K^+ which is known to occur in metabolic acidosis due to the differential response in $[HCO_3^-]_i$ (in metabolic acidosis $[HCO_3^-]_i$ decreases, while in respiratory acidosis, $[HCO_3^-]_i$ increases).

130 What significance do changes in $[HCO_3^-]_i$ have on the cellular levels of K^+ in metabolic alkalosis?

❑ Intracellular $[HCO_3^-]$ increases in metabolic alkalosis and is therefore accompanied by a parallel alteration in the levels of K^+ bound to this anion within the cell. In addition, the K^+-binding capacity of intracellular proteins increases in metabolic alkalosis (see below). Thus, a greater fraction of the total K^+ stores is retained in ICF, at the expense of ECF, favoring the development of hypokalemia.

131 What significance do changes in the anionic equivalency of intracellular proteins have on cellular K^+ levels?

❑ Intracellular alkalosis induces proteins located within ICF to release H^+ and thereby increase their net anionic equivalency. This change in net charge of proteins increases K^+ binding, favoring displacement of K^+ from ECF to ICF. The increase in anionic sites for K^+ binding by intracellular proteins in metabolic alkalosis is additive to that resulting from the higher $[HCO_3^-]_i$.

132 What is the effect of respiratory alkalosis on the K^+-binding capacity of proteins and HCO_3^- within ICF?

❏ In a comparable manner to that described in metabolic alkalosis, the diminished $[H^+]_i$ observed in respiratory alkalosis induces proteins to release H^+, increasing their anionic equivalency. Hence, intracellular K^+ binds to protein sites previously occupied by H^+. The simultaneous presence of a diminished $[HCO_3^-]_i$ will decrease the K^+-binding capacity of this anion. Consequently, the change in K^+ bound by cytosolic HCO_3^- in respiratory alkalosis occurs in association with an opposite change in the K^+-binding properties of the intracellular proteins. The net effect is a milder hypokalemic response to respiratory alkalosis compared to the one elicited by metabolic alkalosis.

133 What is the effect of a decreased K^+-binding capacity of ICF (by proteins, HCO_3^-, and other anions) on $[K^+]_p$?

❏ A decreased K^+-binding capacity of ICF promotes the displacement of cellular K^+ to ECF, resulting in hyperkalemia. The hyperkalemic effect resulting from this mechanism is at its maximum in mineral acid acidoses (HCl or NH_4Cl-induced metabolic acidosis).

134 What is the effect of an increased K^+-binding capacity of ICF (by proteins, HCO_3^-, and other anions) on $[K^+]_p$?

❏ An increased K^+-binding capacity of ICF promotes cellular uptake of K^+ producing hypokalemia. The hypokalemic effect resulting from this mechanism is at its maximum in chloride-sensitive and chloride-resistant metabolic alkalosis.

135 Compare the changes in K^+-binding capacity of ICF in metabolic acid-base disturbances (metabolic acidosis and alkalosis) with those present in respiratory acid-base disturbances (respiratory acidosis and alkalosis).

❏ The net effect in the K^+-binding capacity of ICF that occurs in metabolic acid-base disturbances is substantially greater than the one that occurs in respiratory acid-base disturbances. In metabolic acid-base disturbances there exists an additive change in the K^+-binding properties of intracellular anions, proteins, and HCO_3^-. Conversely, in respiratory acid-base disturbances the change in K^+-binding properties of intracellular proteins is counterbalanced by an opposite change in the K^+-binding properties of intracellular HCO_3^-. Respiratory acidosis increases $[HCO_3^-]_i$ while respiratory alkalosis decreases it; these changes in $[HCO_3^-]_i$ in respiratory acid-base disturbances result from the titration of proteins and of other non-HCO_3^- cell buffers by carbonic acid (H_2CO_3). Therefore, hyperkalemia is more severe in metabolic than in respiratory acidosis. In a comparable manner hypokalemia is more severe in metabolic than respiratory alkalosis. It should be noted that the alterations in $[K^+]_p$ alluded to above are most evident when the change in blood and tissue acidity is acute instead of chronic.

136 Describe the expected change in $[K^+]_p$ in acute acid-base disturbances.

❏ The quantitative relationship between the altered acidity as measured in blood samples and in $[K^+]_e$ is properly expressed as $\Delta[K^+]_p / \Delta pH$. The corresponding values for each acid-base disturbance are as follows:

metabolic acidosis	–7 meq/liter/pH unit
metabolic alkalosis	–4 meq/liter/pH unit
respiratory acidosis	–1 meq/liter/pH unit
respiratory alkalosis	–2 meq/liter/pH unit

137 Is hyperkalemia a constant feature in all types of metabolic acidosis?

❏ No. While metabolic acidosis due to infusion of mineral acid in the experimental animal produces hyperkalemia, organic acid infusion fails to elicit hyperkalemia. Clinical examples of these two models of metabolic acidosis are hyperchloremic metabolic acidosis (e.g., renal tubular acidosis) for the former, and lactic acidosis for the latter.

138 Explain the mechanisms responsible for the disparity in the kalemic response to acute metabolic acidosis due to inorganic (mineral) and organic (e.g., lactic) acids.

❏ During an acid infusion, the H^+ rapidly enters the intracellular compartment independently of the nature of the infused acid (mineral or organic). Maintenance of electroneutrality within the cell requires that the entry of H^+ be accompanied by the entry of an anion (i.e., lactate) or the exit of a cation (i.e., K$^+$) from the cell. The exit of K$^+$ from cells is responsible for the hyperkalemia in mineral acid acidosis, since the anion (i.e., Cl$^-$) accompanying the acid load fails to penetrate the cell, remaining in ECF; thus, to satisfy electroneutrality requirements, K$^+$ exit from the cells is obligatory. On the other hand, organic acids, characteristically have accompanying anions that are easily permeant across cell membranes; therefore, both components of the acid load (anion and cation) enter simultaneously the cell compartment. Consequently, hyperkalemia is not observed in organic acid acidosis since there is no mandatory cellular exit of K$^+$.

139 Is the different cellular permeability of the anion the exclusive mechanism for the disparate kalemic response to acidemia in organic and mineral acid-acidosis?

❏ No. Since the infusion of organic acids (e.g., ketoacids) is accompanied by transient hypokalemia, additional mechanisms besides differences in cellular permeability must be involved. An increase in insulin secretion in response to high plasma levels of organic acids (ketoacids) is responsible for the hypokalemia. Therefore, differences in the cellular permeation of the anion that accompanies the acid load and hormonal influences (infusion of ketoacids stimulates insulin secretion, while HCl-induced metabolic acidosis does not) are involved in the different kalemic response to the infusion of mineral and organic acids.

140 Explain the significance of changes in the intracellular concentration of organic anions in the modulation of K^+ exit from the cells and provide a clinical example.

❑ Lactic acidosis is caused by an increased endogenous production of lactic acid, a decreased disposal of lactate, or a combination of these two processes. This acid-base disturbance results in abnormally high cellular levels of H^+ and organic anions. The intracellular accumulation of H^+ will displace K^+ that was previously electrostatically bound to proteins, phosphate, and bicarbonate. However, the organic anion may act as a sink for the displaced K^+, counterbalancing the effects of the intracellular acidosis on the K^+-binding capacity of cell proteins. In contrast to organic acid-acidosis, the cellular retention of H^+, due to an impairment in renal acid excretion or to infusion of mineral acids, will not only diminish the K^+ binding by proteins, phosphate, and HCO_3^- but will also fail to offer a cellular anion to help with the disposal of the displaced K^+.

141 Explain the effects of changes in the plasma level of other electrolytes on the internal K^+ balance and $[K^+]_p$.

❑ Infusion of NaCl (saline) solutions cause transfer of K^+ from ICF to ECF. Although "dilution acidosis" and increased extracellular osmolality can play a role in this phenomenon, control of these variables does not prevent the translocation of K^+. It is conceivable that changes in the anionic composition of the plasma during the administration of saline solutions (hyperchloremia and/or hypobicarbonatemia) play some role in the exit of cellular K^+.

142 Describe the effect of changes in serum osmolality on $[K^+]_e$.

❑ Hyperosmotic infusions of mannitol or saline can lead to hyperkalemia, independently of their possible effects on acid-base composition. This effect is caused by the rapid extrusion of K^+-rich intracellular water into the extracellular compartment, imposed by extracellular hypertonicity. The same mechanism is responsible for the "paradoxical" hyperkalemia that follows endogenous or exogenous increases in plasma glucose concentration in insulin-deficient diabetics.

143 Explain the determinants of external K^+ balance.

❑ The external K^+ balance is the difference between intake of K^+ and its excretion. A positive external K^+ balance is the consequence of either increased intake, decreased excretion, or a combination of these two mechanisms. Conversely, a negative K^+ balance is the consequence of either decreased intake, increased excretion or a combination of these two mechanisms.

144 Describe the routes of K^+ intake.

❑ Nature has determined that K^+ intake in humans be oral. Parenteral K^+ intake is an alternative route used by physicians when oral intake is either not possible or insufficient. Oral K^+ intake is the physiologic route and allows a more gradual incorporation of K^+ than the parenteral route.

145 Describe the advantages of oral intake in comparison with parenteral intake of K^+.

❑ The pathway of K^+ absorption by the gastrointestinal tract toward its final destination in body tissues mandates an intermediate lodging in the liver; this organ is the largest reservoir of body K^+ other than the skeletal muscle. The intake of K^+ and nutrients by the oral route triggers the release of insulin by the pancreas and activates the sympathetic nervous system; these two responses facilitate the rapid internal disposal of the K^+ load into hepatic and extrahepatic tissues. In contrast, the parenteral intake of K^+ bypasses the liver barrier, therefore exposing the heart and skeletal muscles to a rapid, large increase in $[K^+]_p$ whenever a substantial K^+ intake load is given over a short period of time. The hyperkalemia that can develop in this condition is potentially life-threatening.

146 Describe the routes of K^+ excretion.

❑ The renal excretion of K^+ accounts for approximately 90 to 95% of K^+ intake; the remaining 5 to 10% is excreted in the stools by the gastrointestinal tract. The excretion of K^+ by sweat glands is generally minimal; nevertheless, if excessive sweating occurs, this extrarenal route can cause substantial K^+ losses (as large as 10 to 40 meq/day).

147 Compare fecal K^+ excretion in the normal state and in diarrhea.

❑ Fecal K^+ excretion is very modest in the normal state and only amounts to 8 to 15 meq/day. Potassium loss in diarrheal states increases substantially and can reach as much as 300 meq/day in patients with secretory diarrhea (e.g., non-insulin-secreting islet cell adenoma of the pancreas).

148 What is the major controlling mechanism of renal K^+ excretion?

❏ The major determinant of urinary K^+ excretion is the overall level of body K^+ stores. In states of K^+ excess, renal K^+ excretion rises and in K^+ depletion states, renal K^+ excretion decreases.

149 What is the renal mechanism responsible for the bulk of urinary K^+ excretion?

❏ Potassium secretion by the renal tubule cells is responsible for most if not all of urinary K^+ excretion. The mechanism of K^+ excretion contrasts with that of water and other electrolytes as well as with nonelectrolytes, which are excreted by glomerular filtration and incomplete tubule reabsorption.

150 Describe the nephron segment and the specific cells involved in urinary excretion of K^+.

❏ The principal cells of the distal nephron (late distal tubule and cortical, medullary, and papillary collecting tubule) are responsible for most of the renal excretion of K^+.

151 Describe the various factors involved in the physiologic regulation of urinary K⁺ excretion.

❏ The physiologic modulators of renal K^+ excretion can operate either on the apical membrane (luminal factors) or on the basolateral membrane (peritubular factors) of the K^+-secreting cells of the distal nephron.

152 What are the luminal factors that control renal K^+ excretion?

❏ The luminal factors that control urinary K^+ excretion are the electric profile of the lumen, the urine flow rate, the urine pH, and the urine concentration of the major electrolytes (Na^+, Cl^-).

153 What are the peritubular factors that control urinary K^+ excretion?

❏ The peritubular factors are $[K^+]_e$, dietary K^+ intake (which alters $[K^+]_i$), acid-base status, and hormones (aldosterone, antidiuretic hormone).

154 What is the single most important determinant of K^+ secretion (and therefore renal K^+ excretion) by the distal nephron?

❏ The electric profile of the distal nephron (i.e., degree of electronegativity of the tubule lumen compared with that of the blood) is the most critical determinant of renal K^+ excretion. While the basolateral membrane of the K^+-secreting cell faces the ECF (which has a relatively low $[K^+]$ of ~4 meq/liter), the apical membrane faces the urinary fluid. The net movement of K^+ from the cells to the urinary fluid (K^+ secretion) is the result of a favorable electric gradient (negative lumen potential) that allows K^+ to exit the cell through a conductive pathway located in the apical membrane; thus the urine $[K^+]$ rises to levels that can be severalfold higher than those present in ECF.

155 What is the major determinant of the negative lumen potential in the distal nephron?

❏ The electrically negative potential of the lumen in the distal nephron is due to Na^+ (a cation) reabsorption while its accompanying anion (i.e., Cl^-, HCO_3^-, $HPO_4^=$) is left in the lumen. Therefore, the negative potential generated in the lumen will favor the exit of K^+ from the cell and incorporation to the urine. The lumen negative potential can be as high as -50 mV.

156 Calculate the electrochemical equilibrium of K^+ (E_{K^+}) in distal tubule cells. This calculation will help to explain the contribution of the electric profile of the distal nephron to K^+ secretion.

❏ Let us consider a cytosolic potential difference (PD) of -80 mV, extracellular PD of 0 mV; $[K^+]_e$, 4 meq/liter; and ($[K^+]_i$), 140 meq/liter. We must apply the Nernst equation, as follows:

$$E_{K^+} = 58 \log \frac{[K^+]_e}{[K^+]_i}$$

$$= 58 \log \frac{4}{140}$$

$$= 58 \log 0.029$$
$$= 58 (-1.54)$$
$$= -89 \text{ mV}$$

Thus, K^+ is only slightly above its electrochemical equilibrium, considering that its E_{K^+} is −89 mV and MP is −80 mV.

157 Explain the contribution of the electric profile of the distal nephron to K^+ secretion considering that the lumen PD is −40 mV (an arbitrarily chosen value).

❏ Let us calculate the expected lumen [K^+] that will satisfy the electrochemical equilibrium using the PD gradient across the apical membrane (difference of intracellular PD and luminal PD, named PD_i and PD_{lumen}, respectively) and $[K^+]_i$. The PD gradient is calculated as follows:

$$PD \text{ gradient} = PD_i - PD_{lumen}$$
$$= -80 \text{ mV} - (-40 \text{ mV})$$
$$= -40 \text{ mV}$$

The value obtained (−40 mV) will be inserted into the Nernst equation to replace the term E_{K^+}, since this is the electric potential that will determine the [K^+] in the lumen (urine). The urinary [K^+] is therefore calculated as follows:

$$E_{K^+} = 58 \log \frac{[K^+]_{lumen}}{[K^+]_i}$$

$$-40 = 58 \log \frac{[K^+]_{lumen}}{140}$$

$$40 = 58 \log \frac{140}{[K^+]_{lumen}}$$

$$\log \frac{140}{[K^+]_{lumen}} = \frac{40}{58}$$

$$= 0.69$$

$$\frac{140}{[K^+]_{lumen}} = 10^{0.69}$$

$$= 4.9$$

$$[K^+]_{lumen} = \frac{140}{4.9}$$

$$= 29 \text{ meq/liter}$$

It is therefore apparent that the lumen electronegativity of −40 mV allows for the generation of a urine [K$^+$] seven times higher than [K$^+$]$_e$. If the urine volume remains constant, the higher the urine [K$^+$], the greater will be the urinary K$^+$ excretion.

158 What is the effect of aldosterone and deoxycorticosterone acetate (DOCA) on the electric profile of the distal nephron and on urinary K$^+$ excretion?

❑ Aldosterone and DOCA are two mineralocorticoid agents that increase Na$^+$ reabsorption in the distal nephron, creating a more negative lumen potential, thereby favoring the transfer of K^+ from K^+-secreting cells into the lumen (urine). Therefore, aldosterone and DOCA stimulate renal K^+ excretion.

159 What is the effect of amiloride on PD-mediated stimulation of K^+ secretion resulting from the administration of aldosterone and DOCA?

❑ Amiloride inhibits cellular Na^+ uptake across the apical membrane, thereby reducing the electric driving force that favors K^+ secretion. Thus, the lumen potential is brought toward 0 mV and distal K^+ secretion is substantially diminished.

160 How much will urinary K^+ excretion diminish in the presence of amiloride?

❑ Let us calculate the expected urinary $[K^+]$ using the PD gradient between the intracellular and the urinary compartments ($PD_i - PD_{lumen}$) and $[K^+]_i$, as follows:

$$\begin{aligned} PD\ gradient &= PD_i - PD_{lumen} \\ &= -80\ mV - (0\ mV) \\ &= -80\ mV \end{aligned}$$

The value obtained (-80 mV) will be inserted in the Nernst equation to replace the term E_{K^+}, since this is the electric potential that will determine the $[K^+]$ in the lumen (urine). We must apply the Nernst equation, as follows:

$$E_{K^+} = 58 \log \frac{[K^+]_{lumen}}{[K^+]_i}$$

$$-80 \text{ mV} = 58 \log \frac{[K^+]_{lumen}}{140}$$

$$80 \text{ mV} = 58 \log \frac{140}{[K^+]_{lumen}}$$

$$\log \frac{140}{[K^+]_{lumen}} = \frac{80}{58}$$

$$= 1.38$$

$$\frac{140}{[K^+]_{lumen}} = 10^{1.38}$$

$$= 24$$

$$[K^+]_{lumen} = \frac{140}{24}$$

$$= 6 \text{ meq/liter}$$

Thus, amiloride-induced dissipation of the lumen electronegativity renders the urine [K^+] values comparable to those found in ECF. Potassium excretion is thereby substantially diminished.

---

161 How does urine flow rate modulate renal K^+ excretion?

❏ Since urinary K^+ excretion is determined by the product of urine [K^+] and flow rate, the larger the diuresis, the greater will be the K^+ excretion. It should be apparent that a decrease in urine [K^+] can counterbalance the effect of a higher flow rate if their product remains unchanged or if it decreases. However, a larger diuresis will increase

K^+ secretion by the distal nephron, since the rapid urine flow prevents a build up of luminal $[K^+]$ opposing K^+ secretion, due to a smaller chemical gradient between the cells and the lumen. The increased urine flow observed with the administration of diuretic agents (e.g., acetazolamide, thiazides, furosemide) partly explains the larger urinary K^+ excretion that accompanies the use of these drugs.

162 How does the urine concentration of Na^+ and Cl^- modulate urinary K^+ excretion?

❏ The presence of a high $[Na^+]$ in the lumen of the distal nephron plays a major role in urine K^+ excretion. The increased concentration of Na^+ in the urine observed with the administration of diuretics (i.e., acetazolamide, thiazides, furosemide), other than the K^+-sparing diuretics, partly explains the larger urinary K^+ excretion that accompanies the use of these drugs. When luminal $[Na^+]$ is below 30 meq/liter, K^+ secretion drastically diminishes. The effect of the luminal $[Na^+]$ is explained by its influence on the luminal potential difference (PD). A low urinary $[Na^+]$ prevents the generation of a large negative luminal PD. Contrary to the effect of a low luminal $[Na^+]$, a low luminal $[Cl^-]$ stimulates distal K^+ secretion. The increased kaliuresis observed when the urine $[Cl^-]$ is very low results from stimulation of KCl cotransport across the apical membrane of the distal nephron. This electrically silent K^+-secretory flux might occur through either a symport KCl pathway, or by a parallel electrically coupled Cl^- and K^+ conductance, in the apical membrane of the K^+-secreting cells.

163 How does $[K^+]_e$ modulate urinary K^+ excretion?

❑ A high $[K^+]_e$ increases kaliuresis while a low $[K^+]_e$ decreases it. The effects of $[K^+]_e$ on kaliuresis are mediated by changes in the chemical gradient between blood/distal tubule cells and urine. The $[K^+]_i$ in tubule cells will increase with a high K^+ intake and in hyperkalemia, thereby increasing the K^+ gradient between cells and urine. The opposite is true with a low K^+ intake and in hypokalemia. When a K^+ load is given in association with amiloride treatment, the absolute rate of kaliuresis is very modest. Although not critical for the development of kaliuresis in response to a K^+ load, an unimpaired electric profile as well as unimpaired Na^+ transport mechanisms in the distal nephron (amiloride absent), are responsible for a severalfold rise in renal K^+ excretion above the basal levels observed with a K^+ load.

164 Describe the effects of high dietary K^+ intake on renal K^+ excretion.

❑ A sustained high dietary K^+ intake induces a renal and extrarenal response characterized by an enhanced disposal of the K^+ load. This process, known as "K^+ adaptation", increases the capacity of various nephron segments (cortical collecting tubule, and medullary collecting duct) to secrete K^+ and, therefore, enhances renal K^+ excretion. A high dietary K^+ intake stimulates aldosterone secretion which in turn increases the number and activity of K^+ pumps (Na^+, K^+-ATPase) in the distal nephron. The aldosterone-induced enhancement of the Na^+, K^+-pump and the increased apical conductance of both Na^+ and K^+ in the distal nephron increases kaliuresis, which is a characteristic adaptation observed in individuals on a high K^+ diet.

165 Explain the process of K^+ adaptation of distal nephron and colon that leads to increased K^+ secretion and excretion.

❑ The process of K^+ adaptation of distal nephron and colon is identical to the effect of aldosterone on these tissues. Furthermore, stimulation of aldosterone secretion is most likely responsible for most, if not all, of the adaptive increase in renal and extrarenal K^+ excretion in response to high dietary K^+.

166 Does the adaptation to K^+ loading involve any extrarenal tissue?

❑ Yes. Potassium adaptation, that is, the increased capacity of K^+ excretion by epithelial cells in response to K^+ loading, involves both renal and extrarenal tissues. The extrarenal tissue that participates in K^+ adaptation is the colon (proximal and distal segments) where K^+ secretion can increase from values of 10 meq/day to 20 meq/day or even more in some circumstances (e.g., diarrhea).

167 Describe the effects of aldosterone and DOCA (mineralocorticoid agents) on the renal K^+ excretion.

❑ Mineralocorticosteroids stimulate urinary K^+ excretion. These agents predominantly increase kaliuresis by creating a more favorable electric gradient (a more electronegative lumen) for K^+ secretion. The more electronegative gradient is a consequence of the increased tubule reabsorption of Na^+.

---

168 Explain the effect of aldosterone on the distal nephron and the colon that leads to increased K^+ secretion (and excretion).

❑ The primary effect of aldosterone on these epithelia is the incorporation of new Na^+ and K^+ channels in the apical membrane. The increased activity of the Na^+, K^+-pump with aldosterone may not be a primary effect, but a secondary one, resulting from a higher Na^+ transport due to increased number of apical Na^+ channels. Stimulation of the Na^+, K^+-pump with this hormone is associated with morphologic changes in K^+-secreting cells, which develop an enlarged basolateral membrane area. The increased conductance of apical K^+ channels produces a higher K^+-secretory rate as well as a higher $[K^+]$ in the lumen.

---

169 Compare the effects of mineralocorticosteroids and glucocorticosteroids on renal K^+ excretion.

❑ Body K^+ stores decrease with both types of steroids, yet the glucocorticosteroid-induced K^+ depletion is due to effects on skeletal muscle while the mineralocorticosteroid effects are mostly due to actions on the kidney. Consequently, dietary

NaCl restriction prevents renal K^+ loss due to mineralocorticosteroid excess but does not have a measurable effect in states of glucocorticosteroid excess. In a similar fashion, spironolactone blocks K^+ loss in states of mineralocorticosteroid excess but its does not have an effect on Na^+ and K^+ excretion in states of glucocorticosteroid excess. Finally, $[K^+]_p$ and urinary Na^+ excretion are reduced by mineralocorticosteroid excess but are unchanged (or even increased) in states of glucocorticosteroid excess.

---

170 Provide additional information on the effects of mineralocorticosteroids and glucocorticosteroids on K^+ metabolism.

❏ Mineralocorticosteroids and glucocorticosteroids increase K^+ excretion, but their respective mechanisms of action differ. The glucocorticoid-induced reduction in skeletal muscle mass is partly responsible for the enhanced kaliuresis (negative K^+ and nitrogen balance) observed with these agents. Glucocorticosteroids, in pharmacologic doses, increase K^+ secretion by the colon; aldosterone, in physiologic and pharmacologic doses, increases K^+ secretion in all target tissues including distal nephron, colon, salivary glands, and sweat glands. The kaliuresis, caused by mineralocorticoid action, but not by glucocorticoid action, is associated with Na^+ retention.

171 What changes in electrolyte composition of urine and stool are observed in patients with hyperaldosteronism (primary and secondary)?

❏ Increased Na^+ reabsorption and K^+ secretion in the renal tubule and colon, in response to aldosterone, decreases the Na^+/K^+ ratio in urine and feces.

172 What is the effect of antidiuretic hormone (ADH) on renal K^+ excretion?

❏ ADH enhances renal K^+ excretion. This hormone stimulates the accumulation of K^+ in the renal medulla and results in high $[K^+]$ in medullary interstitial fluid. In addition, ADH increases the permeability of the collecting duct. Consequently, urine flowing down the collecting duct will equilibrate with the high $[K^+]$ in the medullary interstitial fluid resulting in substantial kaliuresis.

173 Describe the effects of a low NaCl diet on urinary K^+ excretion.

❏ A low salt (NaCl) diet decreases delivery of Na^+ to the distal nephron due to increased Na^+ reabsorption in more proximal segments. The diminished natriuresis (diminished Na^+ excretion) impairs K^+ excretion. Aldosterone secretion, however, is stimulated by the ECF volume deficit observed with the low NaCl intake; high aldosterone levels increase K^+ excretion. The opposing effects on K^+ excretion, of reduced distal nephron Na^+ delivery and increased aldosterone levels, counterbalance each other producing a normal renal K^+ excretion.

174 Describe the effects of a high NaCl diet on urinary K^+ excretion.

❑ A high NaCl diet leads to increased NaCl stores and expands ECF volume. The enlarged ECF volume inhibits Na^+ reabsorption in the proximal nephron and allows for a greater delivery of Na^+ to the distal nephron. The increased natriuresis will generate greater kaliuresis. Aldosterone secretion is inhibited, however, by the ECF volume expansion that accompanies the high NaCl diet, an effect that will decrease renal excretion of K^+. The opposing effects on K^+ excretion, of increased distal nephron Na^+ delivery and decreased aldosterone levels, counterbalance each other producing a normal renal K^+ excretion.

175 Describe the effects of water deprivation on renal K^+ excretion.

❑ Water deprivation decreases urine flow in the distal nephron diminishing urinary K^+ excretion. Water deprivation also leads to hypertonicity of body fluids and increases the release of ADH. The stimulation of ADH increases $[K^+]$ in the renal medulla and facilitates equilibration of this compartment with the collecting duct urine. Therefore, high ADH levels increase the urine $[K^+]$ and counterbalance the possible reduction in K^+ excretion due to the antidiuresis (reduced urine volume).

176 Describe the effects of a positive water balance (high water intake) on renal K^+ excretion.

❏ When a positive water balance exists, events opposite to those described in the case of water depletion unfold. The increased urine volume (water diuresis) observed in this condition will augment distal nephron K^+ secretion. Water excess leads to hypotonicity of body fluids and decreases the release of ADH. The absence of antidiuretic hormone will inhibit equilibration of the collecting duct urine with the K^+-rich medulla; additionally, it will diminish the $[K^+]$ in medullary interstitial fluid. The net effect of these opposing processes on urine K^+ excretion is an unchanged renal K^+ excretion.

177 Do changes in water and salt (NaCl) balance significantly alter renal K^+ excretion?

❏ No. We have previously indicated that homeostatic mechanisms allow for the maintenance of renal K^+ excretion in normal individuals exposed to wide fluctuations in water and NaCl intake. Thus, the various mechanisms that modulate renal K^+ excretion can interplay, securing K^+ homeostasis in the presence of an abnormal salt and water balance (of moderate severity).

178 Explain the effects of changes in acid-base status on renal K^+ excretion.

❏ Acute acid-base disorders elicit prompt and substantial changes in renal K^+ excretion. Acute metabolic or respiratory acidosis leads to a sharp decrement in K^+ excretion. In contrast, acute metabolic or respiratory alkalosis leads to a brisk increment in renal K^+ excretion. These phenomena largely reflect changes in the chemical gradient of K^+ between distal tubule cells and lumen; acidosis shift K^+ out of the cells, and alkalosis has the opposite effect.

179 Compare the immediate and sustained response of renal K^+ excretion in acid-base disorders.

❑ The immediate response of renal K^+ excretion in acidosis (respiratory and metabolic) is a decreased kaliuresis, while in alkalosis (respiratory and metabolic), there is an increased kaliuresis. This response is short-lasting. The sustained response in acidosis, respiratory and metabolic, is a greater than normal K^+ excretion leading to a K^+ deficit that is mild in respiratory acidosis and moderate in metabolic acidosis. In respiratory alkalosis, potassium excretion promptly returns to normal so that a significant K^+ deficit does not develop in this condition. In contrast with respiratory alkalosis, a larger K^+ deficit develops in metabolic alkalosis, due to a sizable kaliuresis.

180 Explain the increased kaliuresis observed in respiratory and metabolic acidosis (sustained phase).

❑ The increased K^+ excretion observed in chronic acidosis (respiratory and metabolic) is elicited by an increased delivery of Na^+ to the distal nephron

$$E_{K^+} = 25 \ln \frac{[K^+]_e}{[K^+]_i}$$

$$= 25 \ln \frac{2.5}{139}$$

$$= 25 \ln 0.018$$
$$= 25 \times (-4.0)$$
$$= -100 \text{ mV}$$

43 Apply the Nernst equation to calculate E_{ion} for the distribution of K^+ in frog skeletal muscle using a decimal logarithm to express the ratio of $[K^+]_e/[K^+]_i$. The values for $[K^+]_e$ and $[K^+]_i$ were provided in the previous question.

❑ As above, we insert the values into the equation as follows:

$$E_{K^+} = 58 \log \frac{[K^+]_e}{[K^+]_i}$$

$$= 58 \log \frac{2.5}{139}$$

$$= 58 \log 0.018$$
$$= 58 \times (-1.75)$$
$$= -101 \text{ mV}$$

44 What determines the use of a positive or a negative sign when describing the value of the membrane potential (MP)?

❑ The positive or negative (+ or –) sign given to the MP is determined by the electric potential within the cell compared to that of the ECF. When the MP is measured with intracellular electrodes, the extracellular solution is connected to ground. This is how the zero reference potential is obtained.

45 Compare the equilibrium potential for the distribution of K^+ (E_{K^+}, calculated) with the actual membrane potential (MP, measured) in frog muscle.

❏ The E_{K^+} previously described is -100 mV. The measured MP (or PD) in frog muscle is -92 mV. Consequently, E_{K^+} and MP are very close, with the latter, somewhat smaller (closer to ground).

46 Give a simple interpretation of the discrepancy between E_{K^+} (calculated) and actual MP (measured) in frog muscle.

❏ Since the calculated potential required to maintain the high K^+ concentration (positively charged ion) within the cell is more electronegative than the one actually measured, we can conclude that: (1) the negative PD within the cell is not large enough to explain the measured $[K^+]_i$; (2) mechanisms other than the passive distribution of K^+ (e.g., active pumping of K^+ from ECF to ICF) by electric forces are required to explain the measured $[K^+]_i$; and (3) K^+ within the cell is above its electrochemical equilibrium ($[K^+]_i$ higher than expected).

47 Apply the Nernst equation to calculate E_{ion} for the distribution of Na^+ in frog skeletal muscle with $[Na^+]_e$ and $[Na^+]_i$ 140 meq/liter and 15 meq/liter, respectively.

❑ Let us insert the respective values into our "user friendly" form of the Nernst equation:

$$E_{Na^+} = 25 \ln \frac{[Na^+]_e}{[Na^+]_i}$$

$$= 25 \ln \frac{140}{15}$$

$$= 25 \ln 9.33$$
$$= 25 \times 2.23$$
$$= +56 \text{ mV}$$

48 Apply the Nernst equation to calculate E_{ion} for the distribution of Na^+ in frog skeletal muscle using a decimal logarithm to express the ratio of $[Na^+]_e/[Na^+]_i$. The values for $[Na^+]_e$ and $[Na^+]_i$ were provided in the previous question.

❑ As above, we insert the values into the equation:

$$E_{Na^+} = 58 \log \frac{[Na^+]_e}{[Na^+]_i}$$

$$= 58 \log \frac{140}{15}$$

$$= 58 \log 9.33$$
$$= 58 \times 0.97$$
$$= +56 \text{ mV}$$

49 Compare E_{ion} (calculated) for the distribution of Na^+ (E_{Na^+}) with actual MP (measured) in the frog muscle.

❏ The E_{Na^+} previously described is +56 mV. The MP measured in frog muscle is −92 mV. Thus, the two values differ substantially not only in numerical value but, most importantly, in the sign or polarity.

50 Give a simple interpretation of the discrepancy between E_{Na^+} and actual MP (measured) in frog muscle.

❏ The large discrepancy between calculated and measured MP for Na^+ in frog muscle indicates that: (1) $[Na^+]_i$ is substantially lower than the predicted value, considering the magnitude of the negative cell MP (−92 mV); (2) mechanisms other than passive distribution of Na^+ (e.g., low cell membrane permeability to Na^+, active pumping of Na^+ out of ICF) are required to explain the measured low $[Na^+]_i$; and (3) $[Na^+]_i$ is below its electrochemical equilibrium ($[Na^+]_i$ lower than expected).

51 Calculate the predicted $[Na^+]_i$ that would produce E_{Na^+} identical to MP, considering MP of −92 mV and a $[Na^+]_e$ of 140 meq/liter.

❏ Let us insert the values provided into the Nernst equation considering the hypothetical case in which a MP of −92 mV is identical to E_{Na^+}.

Thus,

$$\text{MP} = 25 \ln \frac{[\text{Na}^+]_e}{[\text{Na}^+]_i}$$

$$-92 = 25 \ln \frac{[\text{Na}^+]_e}{[\text{Na}^+]_i}$$

$$92 = -25 \ln \frac{[\text{Na}^+]_e}{[\text{Na}^+]_i}$$

$$= 25 \ln \frac{[\text{Na}^+]_i}{[\text{Na}^+]_e}$$

$$\ln \frac{[\text{Na}^+]_i}{[\text{Na}^+]_e} = \frac{92}{25}$$

$$= 3.68$$

$$\frac{[\text{Na}^+]_i}{[\text{Na}^+]_e} = 39.6$$

$$\begin{aligned}[]
[\text{Na}^+]_i &= 39.6 \times [\text{Na}]_e \\
&= 39.6 \times 140 \\
&= 5{,}544 \text{ meq/liter}
\end{aligned}$$

It is therefore apparent that $[\text{Na}^+]_i$ of frog muscle of 15 meq/liter is minimal in comparison to the hypothetical $[\text{Na}^+]_i$ of 5,544 meq/liter that would be required if this ion were at electrochemical equilibrium.

52 How does the Gibbs-Donnan rule relate to the electrochemical equilibrium of ions assessed by the Nernst equation?

❏ The Gibbs-Donnan rule predicts the existence of a K^+-high, Na^+ and Cl^--low intracellular concentration in comparison with their extracellular values, influenced by the impermeant intracellular macromolecules. The unique ionic distribution predicted by the Gibbs-Donnan rule will generate a MP (or PD). By contrast, the Nernst equation allows estimation of the electrochemical equilibrium of a particular ion across a membrane, given the ionic concentration in the two compartments.

53 Explain briefly the concept of intracellular Na^+ level being below its electrochemical equilibrium.

❏ The chemical driving force for Na^+ strongly favors its movement from ECF to ICF since its extracellular to intracellular concentration ratio is ~9 to 1. The electric driving force for Na^+ also strongly favors its movement from ECF to ICF since this positively charged ion is driven by the negative electric potential of ICF. The additive effect of the chemical and electric gradients should result in a $[Na^+]_i$ that is much greater than the value actually observed (i.e., a concentration of 5,544 meq/liter compared to one of 15 meq/liter, respectively). We conclude therefore that the Na^+ level within the cell is lower than expected or, in other words, that $[Na^+]_i$ is below its electrochemical equilibrium.

54 How do cells prevent the intracellular accumulation of Na^+ which should result from the favorable chemical and electric gradients described above?

❑ The passive cellular entry (influx) of Na^+ driven by electrochemical forces is counterbalanced by the relatively low cell membrane permeability to Na^+ and by active extrusion of Na^+ by a pump (Na^+, K^+-ATPase) located in the cell membrane. This Na^+ pump spends energy from ATP hydrolysis; the latter is a high energy product of substrate metabolism. Decomposition of ATP to ADP and phosphate, by the pump, is directly coupled to the movement of Na^+ and K^+ across the membrane.

55 Describe briefly the main characteristics of active transport of Na^+ from cells to ECF.

❑ The main characteristics of active Na^+ transport are as follows: (1) the pumping of Na^+ out of cells is coupled to the cellular entry of K^+; thus, this pump is called the Na^+, K^+-pump; (2) the operation of the pump requires energy expenditure from ATP hydrolysis; hence, this pump is known as the Na^+, K^+-ATPase; and (3) the cessation of pump activity resulting from lack of energy (i.e., oxygen, glucose) or the presence of inhibitory drugs results in Na^+ accumulation and K^+ depletion within the cell; the difference in concentration of Na^+ and K^+ between the cellular and extracellular environments diminishes since the ions obey the passive distribution according to electrochemical forces.

56 What substances inhibit the action of Na^+, K^+-ATPase?

❑ The cardiac glycosides (digitalis and related compounds) such as ouabain and strophanthidin are specific inhibitors of the Na^+, K^+-ATPase. Inhibition of the Na^+, K^+-ATPase with these agents has contributed to our understanding of the operation of the pump and its physiological importance. Furthermore, inhibition of the Na^+, K^+-ATPase with cardiac glycosides represents a major therapeutic strategy in the treatment of congestive heart failure and cardiac arrhythmias.

57 Describe other important features of the Na^+, K^+-pump.

❑ The Na^+, K^+-pump is present in all types of animal cells and is a major pathway for cellular Na^+ exit and K^+ entry. The activity of the pump is controlled by the concentration of Na^+ and K^+ in body fluids; an increase in $[Na^+]_i$ and/or an increase in $[K^+]_e$ stimulates ion transport by the pump, and ATP hydrolysis energizes this transport. Conversely, decreasing $[Na^+]_i$ and/or decreasing $[K^+]_e$ slows pump activity. ATP depletion inhibits the Na^+, K^+-ATPase.

58 Describe the stoichiometric relationship of Na^+ transport, K^+ transport, and ATP consumption, by the Na^+, K^+-pump.

❑ The operation of the pump results in the hydrolysis of 1 molecule of ATP for the translocation of three Na^+ out of the cell, and two K^+ into the cell. The above-mentioned stoichiometry is considered fixed under normal conditions and is present in all cells.

59 What are the primary functions of the Na^+, K^+-pump?

❑ The primary functions of the Na^+, K^+-pump include its role in: (1) cell membrane potential; (2) cell volume regulation; (3) maintenance of a low $[Na^+]_i$ and a high $[K^+]_i$ (this feature is closely interrelated to all the primary functions listed); (4) electric excitability; (5) cell incorporation of amino acids, glucose, and other nutrients via cotransport with Na^+; and (6) its role in regulation of intracellular pH via Na^+/H^+ counterexchange.

60 Elaborate further on the regulation of the Na^+, K^+-pump by changes in the concentration of Na^+, K^+, and ATP.

❑ Under physiologic conditions the cellular levels of ATP, $[K^+]_i$, and $[K^+]_e$ are higher than those required for the optimal operation of the pump. Thus, these factors do not play a major regulatory role in its operation. By contrast, the Na^+ concentration for half-maximal activity of the pump is ~30 ± 20 meq/liter, a level that is within the physiologic range of $[Na^+]_i$. The $[Na^+]_i$ is, therefore, an important regulator of the Na^+, K^+-pump.

61 Describe the alteration in cell volume regulation that results from inhibition of the Na$^+$, K$^+$-ATPase (i.e., that one caused by ouabain or anoxia).

❑ Inhibition of the Na$^+$, K$^+$-pump results in loss of intracellular K$^+$ in exchange for extracellular Na$^+$. The K$^+$ diffusion potential across the cell membrane decreases because of diminished [K$^+$]$_i$; consequently, the cell membrane depolarizes, allowing extracellular Cl$^-$ to enter cells. Since the cell-restricted polyanionic macromolecules cannot move to ECF to counterbalance the entry of Cl$^-$, additional Na$^+$ must follow the cellular entry of Cl$^-$ to maintain electroneutrality. Therefore, inhibition of the Na$^+$, K$^+$-pump allows Na$^+$ to enter cells in association with Cl$^-$, in exchange for K$^+$. Cell swelling due to water influx occurs because of increased cellular concentration of osmotically active particles. Certain disorders of the central nervous system are included among the examples of abnormal cell volume regulation that have major clinical relevance. Hypoxic encephalopathy results in brain swelling, and intracranial hypertension develops because of the rigid structure (skull) that surrounds the brain. The intracranial hypertension leads to decreased cerebral blood perfusion, which can produce permanent brain damage or even death.

62 Describe a method to assess the ability of a membrane to allow the movement of ions through it.

❑ The application of a voltage difference (PD or ΔV) to the fluid compartments separated by the membrane will drive ions across the membrane, carrying an electric current (I_{ion}). The ability of the membrane to allow translocation of the ion under these conditions is called conductance of the membrane (G_{ion}). This property is quantified by dividing the measured I_{ion} by the ΔV applied to the membrane. Thus,

$$G_{ion} = \frac{I_{ion}}{\Delta V}$$

$$I_{ion} = \Delta V \times G_{ion}$$

63 Describe another method to assess the ability of a membrane to allow translocation of an ion species from one compartment to another one.

❏ The imposition of a concentration difference (ΔC_{ion}) and measurement of the resulting flux of ions (J_{ion}) allows quantification of the permeability (P_{ion}) of the membrane. Thus,

$$P_{ion} = \frac{J_{ion}}{\Delta C_{ion}}$$

$$J_{ion} = \Delta C_{ion} \times P_{ion}$$

64 Compare the conductance and the permeability of a membrane for a given ion.

❏ Conductance (G_{ion}) involves the transport of ions through channels driven by electric forces. Permeability (P_{ion}) involves the transport of ions through all available mechanisms (including ion channels) driven by a concentration difference. Therefore, the net movement of ions by electroneutral mechanisms (e.g., simultaneous movement of a cation and anion, or simultaneous movement of two ions with the same charge in opposite directions across the membrane) is measured when assessing the permeability of the membrane but not its conductance.

65 Explain further the importance of the Na^+, K^+-pump in the regulation of cell volume.

❑ Intracellular proteins and other organic compounds are, as previously described, negatively charged and cell restricted. Consequently a positive ion, largely K^+, is held within the cell. The K^+-rich cellular compounds promote water movement into the cell by the mechanism of osmosis. This process leads to progressive cell swelling that must be counterbalanced by other mechanisms to prevent cell rupture. The ability of the Na^+, K^+-ATPase to pump out of the cell more ions than the ones it pumps in (three Na^+ exit the cell while two K^+ are pumped into it) produces a net loss of ions from the cell; this process moves water out of the cell by osmosis, counterbalancing the effect of the cell-restricted particles.

66 What is the major determinant of the resting membrane potential (MP) in most cells?

❑ The resting MP is produced mainly by the diffusion of K^+ (a positively charged ion) from ICF to ECF, which generates an electrically negative cell interior. Consequently, MP in the resting cell is considered a K^+ diffusion potential. The cellular exit of K^+ is the result of the large chemical gradient since $[K^+]_i$ and $[K^+]_e$ in skeletal muscle, are 150 meq/liter and 5 meq/liter, respectively.

67 Describe the minimum requirements necessary to generate a diffusion potential.

❏ The minimum requirements necessary to generate a diffusion potential are the presence of ionic gradients (e.g., $\Delta[K^+]$) across a membrane separating two compartments and a selective permeability of the membrane to that particular ion (in the case of K^+ the membrane is impermeable to the counterbalancing anion).

68 Is the K^+ diffusion potential solely responsible for the cell MP observed in the resting state?

❏ No. The cell membrane in the resting state has a defined permeability to several ions and separates two compartments (ICF and ECF) which contain different concentrations of these ions. The role of each individual ion as a determinant of the cell MP is defined by the product of its membrane permeability (P_{ion}) and its concentration gradient (ΔC) across the membrane. Consequently, other ions besides than K^+, such as Cl^- and Na^+, are partially responsible for the overall MP.

69 Describe the relative importance of K^+, Cl^-, and Na^+, in the generation of the skeletal muscle resting MP.

❏ The high permeability of the membrane to K^+ plus the large concentration gradient of this ion across the cell membrane is responsible for the development of the resting MP, which is mostly a K^+ diffusion potential. In contrast with K^+, resting skeletal muscle cells have low membrane permeability to Cl^- and Na^+ (permeability for Cl^- greater than that for Na^+). Therefore, the participation of these ions in the overall resting MP is minimal, despite the very large concentration gradient for Cl^- and Na^+ across the cell membrane.

70 Describe a condition in which Na^+ plays a major role in the generation of MP.

❏ Activation of excitable tissues, including muscle and nerve, results in a sudden increase in membrane permeability to Na^+, leading to a greater conductance for Na^+ than for K^+; consequently a reversal of the relative conductance of these ions, compared with resting conditions, occurs with this process. The increased Na^+ conductance allows immediate entry of Na^+ (positively charged ion) into the cell, reversing the cell negative potential to a positive one. This process is called depolarization and the MP observed in this stage (+35 mV) approaches the Na^+ diffusion potential (+56 mV in frog muscle).

71 Calculate the resting MP (membrane diffusion potential) considering that the cell membrane is permeable to several different ions.

❑ Calculation of the membrane diffusion potential (MP) is performed with a modified version of the Nernst equation known as the Goldman, Hodgkin, Katz equation. The following three factors must be considered when calculating the contribution made by each ion: (1) the concentration of the ion in ICF and ECF; (2) the permeability of the cell membrane to the ion; and (3) the electric charge of the ion, positive or negative. Thus, the membrane diffusion potential (MP) is:

$$MP = 58 \log \frac{P_{K^+}[K^+]_e + P_{Na^+}[Na^+]_e + P_{Cl^-}[Cl^-]_i}{P_{K^+}[K^+]_i + P_{Na^+}[Na^+]_i + P_{Cl^-}[Cl^-]_e}$$

Note that the extracellular concentration of positive ions (Na^+ and K^+) and the intracellular concentration of negative ions (Cl^-), are inserted in the numerator of the equation. The relative permeabilities of skeletal muscle cell membrane in the resting state to K^+, Cl^-, and Na^+ are 100, 2.5, and 1.0, respectively.

72 Compare the concentration of cations in the ICF of some major tissues.

❑ The $[K^+]_i$ is 140 to 160 meq/liter in muscle (skeletal and cardiac), erythrocytes (red blood cells), and liver. The $[Na^+]_i$ is ~10 meq/liter in muscle (skeletal and cardiac), ~20 meq/liter in erythrocytes (red blood cells), and ~30 meq/liter in hepatocytes. The [Mg^{++}] is ~30 meq/liter in liver and in muscle (skeletal and cardiac), and ~5 meq/liter in erythrocytes.

73 Compare the concentration of anions in the ICF of some major tissues.

❑ The [Cl^-]$_i$ is ~5 meq/liter in muscle (skeletal and cardiac), ~20 meq/liter in hepatocytes, and ~80 meq/liter in erythrocytes. The [HCO_3^-]$_i$ is ~10 meq/liter in muscle (skeletal and cardiac), ~15 meq/liter in hepatocytes, and ~20 meq/liter in erythrocytes. The concentration of organic anions is ~90 meq/liter in skeletal muscle, and ~25 meq/liter in erythrocytes. The intracellular concentration of proteins evaluated by their net anionic charge is ~50 meq/liter in skeletal muscle, and ~40 meq/liter in erythrocytes.

74 Compare the intracellular voltage (membrane potential, MP) in some major tissues.

❑ The mean values of intracellular voltage are -90 mV in muscle (skeletal and cardiac), -40 mV in hepatocytes, and -10 mV in erythrocytes.

75 Describe the pathways involved in the exit of K^+ from the cell interior (ICF).

❑ Potassium can exit the cytosol according to the electrochemical gradient either through K^+ channels (K^+Ch) or by K^+/Cl^- cotransport (simulta neous transport of a positive and a negative ion in electroneutral fashion). The K^+/Cl^- cotransport is facilitated by the Cl^- gradient since this ion also accumulates above its electrochemical equilibrium in at least some cells.

76 Describe the pathways involved in the cellular entry of K^+ (from ECF to ICF).

❑ The most important and best studied mechanism for cellular entry of K^+ occurs via the Na^+, K^+-pump. A second K^+-influx pathway involves the obligatory cotransport of one K^+, one Na^+, and two Cl^- across the cell membrane. This pathway, described in several cell types, is electroneutral and is driven by the favorable electrochemical gradient for cellular Na^+ entry.

77 How do ions move across the cell membrane?

❑ Ions move across the cell membrane through proteins that span the lipid bilayer. The basic structure of the cell membrane is a lipid bilayer that charged particles do not easily traverse requiring, therefore, the existence of the more effective ion pathway involving proteins.

78 What are the basic transport mechanisms of K^+ across the cell membrane?

❑ Proteins spanning the lipid bilayer of the cell membrane play a role in the following three mechanisms of K^+ transport: (1) ion channels; (2) cotransport systems; and (3) Na^+, K^+-pump.

79 Describe the basic properties of K^+ channels in the cell membrane.

❏ Cell membrane K^+ channels (K^+Ch) are water-filled pores tunnelled through the structure of specialized proteins connecting ICF to ECF. Potassium channels are selective, allowing the passage of K^+, but excluding other ions. The channels can be in the open or closed state and this process, called gating, is under physiologic regulation.

80 Compare the electric properties of the cell membrane lipid bilayer and those of K^+Ch.

❏ While the lipid bilayer insulates and acts as a capacitor, the channel acts as a conductor of the electric current being transported by K^+. The passage of ions through this selective tunnel (channel) is controlled by electrostatic forces.

81 Compare the conductance (G_{ion}) of the cell membrane to that of the ion channels.

❏ The overall conductance of the plasma membrane at a given time is the summation of the conductance of the individual ion channels in their open state.

82 Describe the electric current (I_{ion}) that develops as a result of ion flux through a channel in its open state.

❏ A channel might carry one picoampere (10^{-12} ampere) of electric current if the applied potential across the channel is 100 mV and the conductance is equal to 10 picosiemens (10 pS). Examination of various channels show a wide range of conductance.

83 What does rectification of an ion channel indicate?

❏ Rectification of an ion channel indicates that its conductance changes with electric potential (voltage).

84 Which cells contain K^+Ch?

❏ Potassium channels are found in all cells whether they are excitable or not. The K^+Ch are the most diverse among all the different kinds of ion channels.

85 Outline the overall role of K^+Ch.

❏ Potassium channels in their open state stabilize the electric MP since they draw it closer to the K^+ equilibrium potential (E_{K^+}). Consequently, in excitable tissues, open K^+Ch move the MP away from the firing potential. Collectively, activated K^+Ch (open state) lower excitability of the cell, set the resting potential, shorten action potential, and inhibit repetitive firing.

86 Describe the various types of K^+Ch according to their physiologic regulators.

❏ The physiologic regulators allow classification of K^+ channels into the following groups: a) ion-activated; b) voltage-gated; and c) receptor-coupled K^+Ch.

87 Give an example of an ion-activated K^+Ch present in excitable tissues (heart, ganglia).

❏ The Na^+-activated K^+Ch increases its conductance when $[Na^+]_i$ rises and decreases its conductance when the $[Na^+]_i$ diminishes. As mentioned earlier, an increased Na^+ conductance of the cell membrane with the resulting movement of Na^+ into the cell is the initial process observed during cell excitation. The entry of Na^+ shifts the negative cell interior toward zero and eventually to a positive electric potential, a process known as depolarization. Activation of K^+Ch by the cellular entry of Na^+ promotes the traffic of K^+ from ICF to ECF driven by its electrochemical gradient (K^+ within the cell is, in this condition, clearly above its electrochemical equilibrium since the negative cell interior holding K^+ within the cell at rest vanishes during cell excitation with the entry of another cation, Na^+). Consequently, K^+ exits the cell, counterbalancing the Na^+-driven depolarization and prompts the process of repolarization (return of MP to its preexcitation level).

88 Name agents that inhibit the conductance of Na^+-activated K^+Ch.

❏ Tetrodotoxin (TTX) and tetraethylammonium (TEA) are blocking agents for Na^+-activated K^+Ch.

89 Name another ion-activated K^+Ch.

❏ The Ca^{++}-activated high conductance K^+Ch is found in many tissues including neurons, smooth and skeletal muscle, and also in exocrine (pancreas, salivary) and endocrine (pituitary) glands. This channel is linked to the processes of cell membrane repolarization and glandular secretion.

90 What other characteristics are present in Ca^{++}-activated, high conductance K^+Ch?

❏ This K^+Ch is activated with depolarization of the cell membrane as well as with increased $[Ca^{++}]_i$. Cell excitation is characterized by an increased $[Ca^{++}]_i$; by contrast, a diminished $[Ca^{++}]_i$ deactivates the channel. Barium (Ba^{++}), tubocurarine, and quinine are known blockers of this channel.

91 Describe some basic aspects of a voltage-activated K^+Ch (a member of the second group of K^+Ch).

❏ The so-called delayed (outward) rectifier is an important K^+Ch found in muscle (skeletal, smooth, and cardiac) and in nervous tissue. The activity of this K^+Ch facilitates repolarization of the cell membrane. Depolarization of the membrane stimulates this channel allowing rapid exit of K^+ from the cell; this counterbalances the electric effect of the entry of Na^+ to the cell. Thus, this K^+Ch shortens action potential, enhances conduction velocity, and facilitates recovery of the cell MP to its resting state. Ba^{++}, TEA, quinine, and strychnine are blockers of this K^+Ch.

92 Name a member of the receptor-coupled K^+Ch (a member of the third group of K^+Ch).

❏ The ATP-sensitive K^+Ch is a major type of K^+Ch found in the beta cells of the endocrine pancreas and myocytes (heart, skeletal, and smooth muscle). The function of this K^+Ch is linked to insulin secretion, maintenance of the resting MP, and protection from anoxic cell death. The channel is activated (opened) and deactivated (closed) by low and high intracellular ATP levels, respectively, as well as by changes in glucose availability. Oral sulfonylureas (such as tolbutamide and glyburide, drugs used in the treatment of type 2 diabetes mellitus), quinine, and TEA are blockers of this ATP-sensitive K^+Ch.

93 Describe the role of ATP-regulated K^+ channels (K^+Ch) in insulin secretion.

❏ The resting membrane potential of beta cells in the pancreas, as well as in most cells, is dependent on the existence of a K^+ diffusion potential (K^+ exiting the cells is responsible for most of the negative electric potential in the cytosol). The closure of the K⁺Ch induces a decrease of the electronegativity of the cytosol, a process known as depolarization. When the plasma glucose level is normal (5 mM), the beta cell is in a polarized state, since sufficient K⁺ conductance in ATP-dependent K⁺Ch occurs (5 to 10% of channel activity is present). Insulin is not released at this normal plasma glucose level. Increasing the glucose level to 10 mM results in glucose uptake by the beta cells, and its subsequent metabolism (glycolysis and Krebs cycle) leads to an increase in intracellular ATP content and/or ATP/ADP ratio. The change in ATP level (and/or ATP/ADP ratio) elicits the closure of ATP-regulated K⁺Ch whose activity, at a 10 mM plasma glucose level, is only 1%; consequently, beta cells depolarize since the exit of K⁺ from the cells is not enough to electrically counterbalance the cellular entry of Na⁺ and Ca⁺⁺. The depolarization of beta cells opens voltage-dependent calcium channels, eliciting an action potential. The rise in [Ca⁺⁺]ᵢ induces the discharge of insulin from secretory granules. The initiation and termination of each secretory pulse of insulin is caused by the closing or opening of relatively few ATP-regulated K⁺Ch in the beta cells of the endocrine pancreas.

94 Describe the proposed linkage between a defect in ATP-regulated K⁺Ch and the development of non-insulin-dependent diabetes mellitus (NIDDM, also called type 2 DM).

❏ It has been proposed that in NIDDM a decreased production of ATP occurs due to impaired glucose metabolism, producing a low ATP/ADP ratio in the beta cells of the pancreas.

Therefore, glucose fails to initiate insulin secretion, since the ATP-regulated K^+Ch are not inhibited because of insufficient ATP levels. The beta cell in diabetes is not depolarized, the voltage-dependent Ca^{++} channels are not opened, $[Ca^{++}]_i$ remains low, and normal secretory pulses of insulin fail to occur.

95 Describe the effect of hypoglycemic sulfonylureas on K^+Ch, and how this action promotes the secretion of insulin.

❏ Sulfonylureas are compounds that have been used in the treatment of diabetes mellitus (type 2) for many years and are known to stimulate insulin secretion. In the absence of increased ATP levels within beta cells, these compounds induce direct closure of ATP-regulated K^+Ch (under physiologic conditions, the rise of ATP within beta cells precedes and triggers insulin secretion). Consequently, in the absence of hyperglycemia and changes in cellular ATP level (and/or ATP/ADP ratio), sulfonylureas bypass the stimulus-secretion coupling of insulin in the beta cells, increasing insulin secretion.

96 What is the threshold membrane potential?

❏ The threshold potential is the unique cell MP that produces a sharp increase in Na^+ conductance, effecting rapid depolarization of the cell (generation of an action potential).

97 Define action potential.

❏ Action potential is a short-lasting reversal of the resting MP in which the cell interior becomes electropositive, contrasting with the negative cell interior found in the quiescent state. This reversal of membrane potential represents an excitatory event that is propagated along muscle cells and nerves.

98 Describe the effect of a low $[K^+]_e$ (similar to a decreased $[K^+]_p$ or hypokalemia) on the resting MP.

❏ Let us consider that $[K^+]_e$ and $[K^+]_i$ are 2 meq/liter and 140 meq/liter, respectively. We insert these values into the Nernst equation as follows:

$$E_{K^+} = 58 \log \frac{[K^+]_e}{[K^+]_i}$$

$$= 58 \log \frac{2}{140}$$

$$= 58 \log 0.014$$
$$= 58 (-1.85)$$
$$= -107 \text{ mV}$$

It is evident that hypokalemia hyperpolarizes (moves intracellular potential toward higher negative values) the cell membrane of muscles and nerves, inhibiting cell activation. Thus, excitability is decreased, since stimuli fail to induce an action potential.

99 Describe the effect of high $[K^+]_e$ (similar to increased $[K^+]_p$ or hyperkalemia) on resting MP.

❏ Let us consider that $[K^+]_e$ and $[K^+]_i$ are 7 meq/liter and 150 meq/liter, respectively. We insert these values into the Nernst equation as follows:

$$E_{K^+} = 58 \log \frac{[K^+]_e}{[K^+]_i}$$

$$= 58 \log \frac{7}{150}$$

$$= 58 \log 0.047$$
$$= 58 \times (-1.33)$$
$$= -77 \text{ mV}$$

It is evident that hyperkalemia depolarizes (moves intracellular potential toward zero) the cell MP of muscle and nerve, facilitating cell activation of these excitable tissues. Yet, after excitation, hyperkalemia prevents repolarization rendering a tissue that is no longer excitable. It should be apparent that both hyperkalemia and hypokalemia lead, eventually, to paralysis or lack of functioning of excitable tissues.

100 Are changes in excitability mediated solely by alterations in resting MP?

❏ No. Tissue excitability can be modified by alterations in the threshold potential as well as in the resting potential.

101 Which ions modify the threshold membrane potential?

❏ Calcium ions (Ca^{++}) are the ones which primarily alter the threshold MP, contrasting with K^+, which primarily affect the resting MP.

102 What is the effect of a high extracellular concentration of ionized calcium (hypercalcemia) on threshold potential?

❏ Hypercalcemia displaces the threshold MP toward zero (the threshold potential is at a less negative intracellular potential), diminishing tissue excitability. Consequently, hypercalcemia increases the difference between resting potential and threshold potential, therefore, stronger stimuli are required to elicit the greater depolarization needed to induce cell excitation.

103 What is the effect of a low extracellular concentration of ionized calcium (hypocalcemia) on threshold potential?

❏ A low plasma concentration of ionized calcium (hypocalcemia) displaces threshold MP, moving it from zero toward a more negative intracellular potential. Consequently, the difference between resting and threshold potential is reduced, requiring stimuli of less intensity to elicit a response. Therefore, tissue excitability is increased with hypocalcemia and decreased with hypercalcemia.

104 What would be the effect of simultaneous changes in resting and threshold potential on tissue excitability?

❏ Simultaneous changes in resting and threshold MP might have additive or counterbalancing effects on tissue excitability. Additive effects might appear when both resting and threshold MP move toward each other. Counterbalancing effects might occur when both the resting and the threshold MP move away from each other.

Mechanisms of K⁺ Homeostasis

105 What is the normal $[K^+]_p$?

❑ Normal $[K^+]_p$ ranges between 3.5 meq/liter and 5.0 meq/liter.

106 Describe the differences, if any, between serum and plasma $[K^+]$.

❑ Serum is obtained after spontaneous clotting of a blood sample while plasma requires blood drawing with an anticoagulant (heparin) and subsequent centrifugation to separate the blood components (red cells and plasma). The serum $[K^+]$ is slightly but consistently higher than the plasma $[K^+]$, with that difference consisting of ~0.2 meq/liter. The higher $[K^+]$ in the serum is caused by the release of K^+ from red cells. Since the difference between serum and plasma $[K^+]$ is relatively small, the terms serum and plasma are used interchangeably, in clinical practice as well as in this book, when referring to K^+ levels.

107 What are the major determinants of plasma K^+ concentration ($[K^+]_p$), or more broadly, of $[K^+]_e$?

❑ The major determinants of $[K^+]_e$ are the level of total body K^+ stores (normal, low, or high) and the relative distribution of K^+ between ECF and ICF.

108 In which way is the comparatively small extracellular K^+ content (2% of total body K^+ stores) significant in the regulation of $[K^+]_e$?

❏ Since the ratio of $[K^+]_e / [K^+]_i$ is responsible for most of the biological effects of K^+, it is of critical importance that this ratio remains constant. The very large intracellular K^+ stores are not substantially altered in response to physiologic loading and unloading of K^+. By contrast, the tiny extracellular K^+ stores are very susceptible to changes in total body K^+ content. Consequently, physiologic control mechanisms to prevent wide fluctuations only in $[K^+]_e$ (differing from $[K^+]_i$), have developed.

109 What determines the level of total body K^+ stores?

❏ The total body K^+ stores are established by the external K^+ balance, which is, in turn, determined by the difference between K^+ intake and excretion.

110 What are the internal K^+ balance and the external K^+ balance?

❏ The internal K^+ balance refers to the control mechanisms for the distribution of total body K^+ stores between ICF and ECF. Thus, internal K^+ balance refers to the internal exchanges of K^+. On the other hand, external K^+ balance refers to the determinants of total body K^+ stores, without any consideration for the distribution of K^+ among body compartments.

which enhances kaliuresis, overriding the expected renal K^+ retention due to acidosis.

181 Explain the increased kaliuresis leading to a large K^+ depletion in metabolic alkalosis.

❑ The increased $NaHCO_3$ excretion observed, at least intermittently, in the course of metabolic alkalosis produces an alkaline urine, rich in a relatively non-reabsorbable anion (HCO_3^-); both factors (change in pH and presence of HCO_3^-) enhance kaliuresis. The increased plasma aldosterone levels observed in gastric alkalosis (due to volume depletion) as well as in the various forms of chloride-resistant metabolic alkalosis, induce enhanced Na^+ reabsorption and K^+ secretion in the distal nephron.

182 Define fractional K^+ excretion by the kidney.

❑ Fractional K^+ excretion (FE_{K^+}) is the percent of K^+ filtered load that is excreted in the urine. Assuming that K^+ intake is ~75 meq/day, and that the excretion is mostly urinary, ~70 meq of K^+ will be excreted daily by the kidney. The total GFR in a normal individual in 24 hours is ~180 liters, which represents the product of 125 ml/min GFR and 1,440 min (24 hours). The daily K^+ filtered load is ~720 meq which represents the product of 4 meq/liter $[K^+]_e$ and 180 liters (GFR). Consequently, the ratio of urinary K^+ excretion (i.e., 70 meq/day) to the filtered load of K^+ (i.e., 720 meq/day) is ~10%. The FE_{K^+} in normal individuals, therefore, is ~10%.

183 What is the effect of an increased K⁺ load on FE_{K^+}?

❏ An increased K⁺ load triggers K⁺ adaptation that increases FE_{K^+}. Let us assume that dietary K⁺ in a normal individual increases four times over his usual intake (300 meq/day). Since neither GFR nor $[K^+]_p$ substantially change, urinary K⁺ excretion increases fourfold. Consequently, calculated FE_{K^+} is ~40% (FE_{K^+} is ~10% with a standard K⁺ dietary intake).

184 What is the effect of renal insufficiency on FE_{K^+}?

❏ FE_{K^+} increases in renal insufficiency allowing these patients to remain normokalemic (normal $[K^+]_p$) in spite of a major reduction of GFR. Since $[K^+]_p$ remains within normal limits in patients with only 25% of overall renal function (GFR) and K⁺ intake remains in the normal range, a fourfold increase in FE_{K^+} must be present. Since the calculated FE_{K^+} in normal individuals amounts to ~10%, the estimated FE_{K^+} in a patient with this degree of renal insufficiency is ~40%.

185 Compare the risk for the development of hyperkalemia in response to high K⁺ intake of normal individuals and those with only 25% of overall renal function.

❏ Normal individuals will increase FE_{K^+} in response to K⁺ loading; if K⁺ intake increases fourfold (300 meq/day), FE_{K^+} will also increase fourfold (from 10% to 40%). Patients with a severe reduction of GFR (i.e., 25% of normal GFR) maintain K⁺ homeostasis while receiving standard K⁺ intake by increasing to the maximum their capacity to excrete K^+ (FE_{K^+} increases to 40%). A higher than normal K^+ intake given to these patients will result in hyperkalemia since they cannot increase further their renal K^+ excretion.

186 How much does FE_{K^+} increase in response to K^+ loading in normal individuals and in patients with renal insufficiency?

❏ FE_{K^+} can increase to ~50% in both groups of individuals. Levels of FE_{K^+} even higher than 50%, occasionally exceeding 100%, have been reported but should be considered the exception rather than the rule.

187 What is the purpose of having several control mechanisms aimed at achieving a single goal, i.e., preventing a major rise or a major fall in $[K^+]_p$?

❏ Components that are critically important in the operation of a machine, biologic or otherwise, require the presence of multiple "back up" mechanisms to insure their proper and continuous functioning.

The emergency escape door of an aircraft, for example, has a manual backup mechanism to secure its proper operation whenever the automatic process fails. Another example is the presence of overlapping systems to guide the space shuttle, permitting continued operation of the spacecraft should one of the systems fail. Since a major change in $[K^+]_p$ can have lethal consequences in biological systems, multiple mechanisms are available to prevent major fluctuations in the level of this vital constituent.

188 Describe the redundant mechanisms that participate in preventing a lethal decrease in $[K^+]_p$.

❏ Whenever the risk of development of a major decrease in $[K^+]_p$ arises, internal and external mechanisms of K^+ conservation are recruited (contrary to the mechanisms for K^+ disposal that are recruited in hyperkalemia). Redistribution of internal K^+ stores occurs as a result of cellular K^+ exit to ECF through K^+ channels since ~98% of K^+ stores are located in ICF. The defense mechanisms of hypokalemia are facilitated by inhibition of insulin secretion and stimulation of alpha adrenoreceptors in response to hypokalemia. Urinary K^+ excretion decreases to < 10 meq/day since the distal nephron becomes a K^+-reabsorbing epithelium in response to K^+ depletion, reversing its usual role of a K^+-secreting epithelium. Hypokalemia decreases aldosterone secretion, thereby inhibiting renal K^+ excretion.

189 Describe the redundant mechanisms that participate in preventing a lethal rise in $[K^+]_p$ in individuals fed a K^+-rich diet.

❏ Considering that the amount of K^+ absorbed from a single meal can be as large as the total extracellular K^+ content, in the absence of an effective mechanism of K^+ disposal, doubling of the $[K^+]_p$ might occur with such K^+ intake. The so-called internal and external K^+ disposal mechanisms, each of which has various components, are immediately recruited. The oral intake of nutrients stimulates insulin release which facilitates tissue K^+ uptake. Feeding also activates the sympathetic system that facilitates K^+ disposal due to beta adrenergic-mediated mechanisms. The rise of $[K^+]_p$ produced by oral intake promotes the cellular entry of K^+ as well as its urine excretion. In addition, the increased $[K^+]_p$ due to oral K^+ loading promotes aldosterone secretion which enhances the urinary disposal of K^+.

Clinical Disorders

1. Hypokalemia

190 Define hypokalemia.

❏ Hypokalemia is the condition where the $[K^+]_p$ is below 3.5 meq/liter.

191 Provide a broad description of the most salient effects of hypokalemia.

❏ Hypokalemia impairs multiple body functions that are conveniently classified into the following categories: (1) cardiovascular; (2) neuromuscular; (3) renal; and (4) endocrine/metabolic.

192 What are the effects of hypokalemia on cardiac excitability?

❏ The effects of hypokalemia on cardiac excitability are generally opposite to those described for hyperkalemia. Consequently, hypokalemia alters cardiac excitability by hyperpolarizing the resting MP. Hypokalemia also decreases the cell membrane permeability to K^+ ions, prolonging the duration of the action potential, since cellular exit of K^+ is responsible for returning the depolarized cell membrane to its resting state. Furthermore, pacemaker activity is increased since the spontaneous depolarization characteristic of pacemaker cells is stimulated.

193 Describe further the cardiovascular effects of hypokalemia.

❏ The cardiac effects include an increased risk of digitalis toxicity, supraventricular tachyarrhythmias (atrial, junctional), ventricular arrhythmias (isolated beats, bigeminal rhythm, tachycardia, or fibrillation), other characteristic ECG changes (flat or inverted T waves, depressed ST segments, prominent U waves, and a prolonged "QU" interval that resembles the prolonged "QT" interval characteristic of hypocalcemia), and structural myocardial damage (myocardial necrosis). The vascular effect of hypokalemia is constriction of arterioles (resistance vessels).

[margin note: immediately follow T waves]

194 What are the effects of K^+ on vascular resistance of arterioles?

❏ Potassium depletion causes arteriolar constriction in several vascular beds, while K^+ administration has vasodilatory effects. Hypokalemia due to K^+ deficit can induce constriction of coronary and cerebral vessels, possibly leading to ischemic damage of these tissues.

195 Describe the linkage between systemic blood pressure and the level of body K^+ stores.

❏ Increased systemic vascular resistance, producing hypertension, has been postulated to occur in states of K^+ depletion. Potassium repletion can decrease peripheral vascular resistance and diminish systemic blood pressure.

196 Can dietary K⁺ intake play a role in the incidence/severity/complications of hypertension, in different populations?

❏ A lower dietary K⁺ intake of African-Americans, compared to that of European-Americans in the United States can partly explain the higher incidence/severity/complications of hypertension in the former population. Socioeconomical and cultural factors have been invoked as the reasons for the lower K⁺ intake in African-Americans.

197 Describe the neuromuscular effects of hypokalemia.

❏ Hypokalemia impairs skeletal muscle function resulting in weakness, cramps, myalgias, myolysis, paresis, and paralysis. The skeletal muscle dysfunction can involve the extremities and the torso, including the respiratory muscles. Hypokalemia also impairs the function of smooth muscle in the gastrointestinal (e.g., gastric atony, ileus) and urinary (bladder atony, urinary retention) systems.

[handwritten annotation: INCOMPLETE PARALYSIS]

198 Describe the effect of severe K⁺ depletion on the electrolyte composition of skeletal muscle.

❏ Severe K⁺ depletion generates a major reduction in the K⁺ content of cells. Maintenance of cell volume requires that K⁺ loss be replaced by a comparable increment in another cation (e.g., Na⁺). Thus, the sum of the $[Na^+]_i$ and $[K^+]_i$ during K⁺ depletion is similar to that found in the normal state. The $[Cl^-]_i$ is not altered in K⁺ depletion. These electrolyte changes are observed whether K^+ depletion is due to a low K^+ diet or to mineralocorticosteroid excess.

199 What are the major abnormalities of skeletal muscle in states of K^+ deficiency?

❏ The major abnormalities observed in skeletal muscle with K^+ depletion are: (1) a deranged electrochemical profile; (2) diminished glycogen synthesis and storage; (3) decreased muscle blood flow; and (4) necrosis, also called rhabdomyolysis.

200 What is the electric profile of skeletal muscle in the early phase of K^+ deficiency, compared to its late phase?

❏ The predicted resting membrane potential (MP) of skeletal muscle in both early and late phases of K^+ depletion in man is that of a more negative cell interior or hyperpolarization; this predicted electric profile is due to a greater decrement in $[K^+]_e$ with respect to $[K^+]_i$, so that the ratio of $[K^+]_e / [K^+]_i$ decreases. In the acute phase, the actual resting MP agrees with the predicted value, so in this phase, the muscle is hyperpolarized. By contrast, actual resting MP in the late phase of K^+ depletion demonstrates that the muscle is depolarized (cell interior less negative than in normal conditions). This electric profile is best explained by a decrease in the K^+ conductance of skeletal muscle cell membrane. Measured resting potentials in normal muscle, in early phase, and late phase of K^+ depletion are -90 mV, -94 mV, and -55 mV, respectively.

201 What is the $[K^+]_p$ in hypokalemic states in which clinical manifestations of skeletal muscle dysfunction are observed?

❑ While patients with $[K^+]_p > 3$ meq/liter are usually asymptomatic, muscle weakness, increased fatigability, and other symptoms develop when $[K^+]_p$ decreases below this level. Increased serum levels of muscle enzymes (creatine phosphokinase or CPK, serum amino-transferase or SAT that was previously designated SGOT, and aldolase) occur when $[K^+]_p$ is < 2.5 meq/liter, and severe rhabdomyolysis can be observed with $[K^+]_p < 2.0$ meq/liter.

202 Describe the renal effects of hypokalemia.

❑ Hypokalemia decreases the ability to optimally concentrate urine, leading to polyuria and polydipsia (nephrogenic diabetes insipidus). Sodium retention and edema formation is also observed in K^+ depletion. Metabolic alkalosis induced by an increased production and excretion of urinary ammonium (NH_4^+) in association with increased renal HCO_3^- reabsorption can also occur in K^+ depletion.

203 Describe the endocrine/metabolic effects of hypokalemia.

❑ Potassium depletion induces a negative nitrogen balance, causing growth retardation.

Hypokalemia reduces insulin secretion, alters carbohydrate metabolism, decreases glycogen synthesis and content in skeletal muscle, induces glucose intolerance, and aggravates diabetes mellitus. Hyperlipidemia often accompanies hyperglycemia in K^+ depletion. Hypokalemia acting directly on the adrenal gland decreases aldosterone secretion, while hyperkalemia stimulates it; however, K^+ depletion increases plasma renin activity.

Potassium depletion predisposes susceptible individuals to hepatic encephalopathy and hepatic coma.

204. What is the effect of K^+ depletion on the activity of alpha adrenoreceptors?

☐ Potassium depletion activates alpha adrenoreceptors which inhibit Na^+, K^+-ATPase in skeletal muscle. The decreased activity of the Na^+, K^+-pump tends to decrease $[K^+]_i$ with respect to $[K^+]_e$, as well as increase $[Na^+]_i$.

205 Describe the general mechanisms leading to hypokalemia.

☐ Hypokalemia ($[K^+]_p$ < 3.5 meq/liter) can result from redistribution or depletion of K^+ stores. The hypokalemia that results from redistribution is caused by cellular uptake of K^+ that shifts K^+ out of ECF; K^+ redistribution can occur simultaneously with K^+ depletion so that the two processes leading to hypokalemia can have additive effects. The hypokalemia observed with K^+ depletion is attributable to a reduction in the K^+ content of all body fluids.

206 Name the most significant causes of hypokalemia due to K^+ redistribution.

❏ Potassium redistribution leading to hypokalemia is observed with:

1. sympathetic activation (release of catecholamines at nerve terminals and/or to the circulation);

2. exogenous source of beta-adrenergic agonists;

3. insulin effects due to endogenous release or insulin therapy;

4. alkalemia of respiratory or metabolic nature;

5. barium poisoning due to ingestion of soluble barium salts;

6. potassium uptake by red cell precursors (response to vitamin B_{12}, or folate therapy of megaloblastic anemia).

207 Describe the hypokalemia due to activation of beta-2 adrenergic receptors.

❏ Activation of beta-2 adrenergic receptors is responsible for the so called "stress hypokalemia". The endogenous release of catecholamines due to stress (physiological conditions and disease states), or an exogenous source of beta-2 adrenergic agonists (e.g., therapy of obstructive airway disease) stimulates Na^+, K^+-ATPase in skeletal muscle and other tissues, and results in hypokalemia.

208 Explain the mechanism of hypokalemia in patients with prolonged hypothermia.

❑ Hypothermia can be associated with either hypokalemia or hyperkalemia. Hypokalemia in prolonged hypothermia is due to tissue uptake of K^+ secondary to activation of beta-2 adrenergic receptors. Correction of hypothermia returns the decreased $[K^+]_p$ to normal values. Occasionally hyperkalemia occurs in prolonged hypothermia; the presence of hyperkalemia in hypothermic patients indicates profound cellular dysfunction and almost certain death.

209 Describe widely used drugs, excluding beta-2 agonists and insulin, that cause hypokalemia from K^+ redistribution.

❑ Xanthines are drugs widely used in the treatment of reactive airways due to various diseases, including bronchial asthma, "cardiac asthma", and chronic obstructive pulmonary disease (COPD). These drugs mimic the effects of beta-2 agonists on $[K^+]_p$. Both drugs stimulate Na^+, K^+-ATPase as a result of increased cyclic AMP levels. While beta-2 adrenergic agonists enhance production of cyclic AMP, xanthines inhibit degradation of this compound.

210 Describe the clinical settings in which insulin causes hypokalemia.

❑ The endogenous release of insulin, in response to a large glucose intake (oral or parenteral), can induce clinically significant hypokalemia. The exogenous administration of insulin, in the treatment of diabetes mellitus, can also produce hypokalemia. Insulin administration in the therapy of diabetic ketoacidosis, a condition in which K^+ depletion is usually present, can result in profound and symptomatic hypokalemia.

211 Describe the clinical settings in which alkalemia induces hypokalemia.

❑ Alkalemia due to either respiratory or metabolic alkalosis is associated with hypokalemia. When the alkalemia is due to metabolic alkalosis, the hypokalemia is substantially more severe.

212 Explain the mechanism and describe the clinical setting for barium-induced hypokalemia.

❑ Ingestion of soluble salts of barium (Ba^{++}) can produce profound hypokalemia. Once absorbed from the gastrointestinal tract, Ba^{++} is distributed throughout the body; this compound competitively blocks K^+ channels (K^+Ch) in cell membranes but leaves Na^+, K^+-ATPase intact. Consequently, the cellular efflux of K^+ through channels is interrupted but the ATP-dependent cellular K^+ uptake remains active, leading to hypokalemia. Blockade of K^+Ch induces depolarization of skeletal muscle and subsequent paralysis. Displacement of Ba^{++} from the K^+Ch is accomplished by K^+ administration. The oral and rectal administration of Ba^{++} salts, commonly used for the performance of radiologic diagnostic evaluation of the gastrointestinal tract, are safe since the Ba^{++} compounds used are insoluble.

---

213 Describe the mechanism of hypokalemia observed in the treatment of megaloblastic anemia.

❑ Maturation of red cell precursors in response to B_{12} and folate administration in the therapy of megaloblastic anemia result in hypokalemia due to K^+ uptake by the bone marrow. An increased oral K^+ intake can effectively counterbalance the hypokalemic action of folate and B_{12} therapy.

---

214 What explains the presence of hypokalemia or hyperkalemia in the syndromes of periodic paralysis?

❑ The attacks of paresis/paralysis in hypokalemic periodic paralysis are initiated by an excessive K^+ uptake by skeletal muscle, which explains the development of hypokalemia. In a comparable manner, the attacks of paresis/paralysis in hyperkalemic periodic paralysis are initiated by an excessive release of K^+ by skeletal muscle, which is responsible for the development of hyperkalemia.

---

215 Where is the defect responsible for the episodes of periodic paralysis located?

❑ Since propagation of action potentials along nerves is normal while the muscles fail to contract with either direct or nerve stimulation, the primary defect is restricted to the skeletal muscle itself.

216 Name the major syndromes of hypokalemic periodic paralysis.

❑ The main forms of hypokalemic periodic paralysis are: (1) primary familial hypokalemic periodic paralysis; (2) secondary hypokalemic paralysis; (3) thyrotoxic periodic paralysis; and (4) barium-induced periodic paralysis.

217 Describe the main characteristics of primary familial hypokalemic periodic paralysis.

❑ The main clinical features of familial hypokalemic periodic paralysis are similar to those previously described in all forms of hypokalemic paralysis. Its specific features are its inheritance as an autosomal dominant trait, a higher incidence in males (a male/female ratio of 4:1), and a first attack occurring by age 20. Since a meal rich in carbohydrates is known to precipitate attacks of paralysis, the diagnosis of this disease can be confirmed with a protocol that combines glucose and insulin administration. This evaluation by provocative tests is potentially risky because profound hypokalemia and paralysis can develop, so it should be performed by an expert to ensure safety. The main laboratory datum supporting the diagnosis of familial hypokalemic paralysis is the demonstration of normal $[K^+]_p$ between attacks of paralysis.

218 What are the main features of secondary hypokalemic paralysis (periodic and non-periodic)?

❏ The main characteristics of patients with secondary hypokalemic paralysis are the presence of severe K^+ depletion, persistent hypokalemia (contrary to the episodic hypokalemia that accompanies paralytic attacks in the primary/familial form of the disease), and persistent generalized weakness, with or without attacks of skeletal muscle paralysis. Contrasting with the primary/familial form of this disease, the secondary form develops in patients at any age, $[K^+]_p$ is always low, and associated conditions leading to severe K^+ depletion are present (e.g., excessive renal or extrarenal K^+ loss).

219 Describe the Ba^{++}-induced skeletal muscle paralysis.

❏ Barium intoxication leads to hypokalemia as a result of competitive inhibition of K^+ channels in the cell membranes; therefore, K^+ exit from the cell interior is inhibited. The reduction in K^+ conductance depolarizes muscle cell membranes which results in depressed excitability and inactivation of Na^+ channels. The clinical syndrome of Ba^{++}-induced paralysis is due to a combination of hypokalemia and the direct effect of Ba^{++} on K^+ channels of skeletal muscle. This unusual syndrome results from accidental or intentional ingestion of soluble Ba^{++} salts and is characterized by gastrointestinal (vomiting and diarrhea due to hemorrhagic gastroenteritis), cardiac (arrhythmias), and neuromuscular (convulsions, muscle twitching, and muscle paralysis) manifestations. Potassium loss from the gastrointestinal tract and the transfer of K^+ to the intracellular compartment is responsible for the development of Ba^{++}-induced hypokalemia. Administration of K^+ helps to reverse the paralysis by displacing Ba^{++} from the K^+ channels. Contamination of food products by Ba^{++} salts has produced outbreaks of Ba^{++}-induced paralysis.

220 Describe the clinical features in patients with periodic hypokalemic paralysis.

❏ The main characteristics in the syndrome of periodic paralysis are:

1. muscle strength is normal at the onset of the disease and between attacks of paralysis; yet, permanent weakness may develop thereafter;

2. attacks of paresis/paralysis last from hours to days, they are usually generalized and involve upper and lower extremities; the tendon reflexes of all extremities diminish or disappear; the muscles served by cranial nerves, as well as the respiratory muscles are usually spared; and

3. paresis or paralysis commonly develops in the rest period immediately following vigorous exercise; periodic paresis/paralysis can be prevented by exercise of mild intensity. Cold exposure can precipitate paralytic attacks.

221 Name the most significant causes of hypokalemia due to K^+ depletion.

❏ Potassium depletion can occur when dietary K^+ intake is very low and therefore fails to counterbalance the obligatory K^+ losses. Potassium depletion can also occur if K^+ losses are abnormally high and occur in association with a normal dietary K^+ intake. Potassium losses may be renal or extrarenal in origin.

222 Describe the pathogenesis of K^+ depletion in relation to nitrogen balance.

❏ Potassium depletion is always associated with a reduction of intracellular K^+ stores since 98% of body K^+ is located within cells. Potassium depletion can occur: (1) associated with tissue loss (particularly muscle), condition referred to as "pseudodepletion" of K^+ stores because tissue K^+ levels are within normal limits; and (2) not associated with tissue loss, referred to as true K^+ depletion because tissue K^+ levels are abnormally low. The K^+/N ratio (potassium/nitrogen), that in normal tissues is ~3, remains unchanged in "pseudo K^+-depletion" and decreases in true K^+ depletion. This classification of K^+ depletion has some theoretical value but it does not have practical application since the K^+/N ratio is unknown in most clinical conditions. Furthermore, K^+ depletion itself induces a negative nitrogen balance and prevents proper nitrogen anabolism; therefore all forms of K^+ depletion, if severe enough, will impair nitrogen balance.

223 Describe a clinical syndrome of K^+ depletion with normal K^+/N ratio (also called "pseudo K^+-depletion").

❏ Adrenal hypersecretion (mineralocorticosteroids and glucocorticosteroids) or the exogenous administration of ACTH (adrenocorticotropin), in the presence of a low NaCl intake, can result in a negative balance of nitrogen and K^+ of similar degrees, so that the K^+/N ratio remains unchanged. This situation is explained by the effect of adrenocortical hormones on skeletal muscle. If a high NaCl intake accompanies the increased effects of adrenocortical hormones (of endogenous or exogenous origin), excessive kaliuresis will develop as a result of increased delivery of NaCl to the distal nephron and the renal effects of mineralocorticosteroids; thus, K^+ loss will be greater than nitrogen loss and a state of "true" K^+-depletion will now develop.

224 Compare the alteration in K^+ content in ICF and ECF in the presence of total body K^+ deficit.

❏ Total body K^+ deficit results in a greater absolute reduction of K^+ content in ICF than ECF. Nevertheless, the percent deviation in K^+ content is considerably smaller in ICF than ECF. In a similar fashion, the decrease in $[K^+]_i$ with K^+ depletion is significantly smaller than the decrease in $[K^+]_p$. Consequently, the chemical gradient of K^+ efflux from the cell increases in hypokalemic states, resulting in hyperpolarization of the cell membrane (the cell interior becomes more electronegative).

225 Describe the relationship between $[K^+]_p$ and degree of K^+ deficit.

❏ A linear relationship having a slope of 0.3 meq/liter per 100 meq of K^+ deficit ($\Delta[K^+]_p / \Delta K^+$ stores) has been described in patients with K^+ depletion in the absence of redistribution of K^+ stores. According to this relationship, a K^+ depletion of ~10% of total body K^+ stores (350 meq) produces a decrease in $[K^+]_p$ of ~1 meq/liter.

226 Describe the causes of hypokalemia due to K^+ deficit that are primarily due to poor dietary K^+ intake.

❏ A low K^+ intake is characteristically observed in patients ingesting a "tea and toast" diet, alcoholics, clay eaters (geophagia), and in anorexia nervosa.

227 In what clinical setting is hypokalemia due to "tea and toast" diet observed?

❏ Elderly individuals with significant economic constraints, especially those who are edentulous, sometimes choose an almost exclusive "tea and toast" diet to satisfy dietary requirements. While the caloric intake can be adequate due to breakdown of ingested starch, K^+ intake is minimal and leads to hypokalemia due to K^+ depletion.

228 Explain the determinants of the hypokalemia observed in alcoholics.

❏ Hypokalemia is frequently observed in alcoholics and its pathogenesis is multifactorial. Redistribution as well as K$^+$ depletion are responsible for the development of hypokalemia in these patients. A catecholamine surge is observed in alcohol intoxication especially with the syndrome of delirium tremens. Potassium depletion is caused by poor dietary K$^+$ intake, gastrointestinal losses (i.e., vomiting, diarrhea, malabsorption) and excessive renal K$^+$ loss. Renal K$^+$ excretion can be increased in alcoholics due to the intermittent excretion of urinary HCO_3^-, the simultaneous presence of secondary hyperaldosteronism (i.e., alcoholic liver disease), and severe magnesium depletion.

229 How does clay ingestion lead to hypokalemia?

❏ Clay ingestion, also known as geophagia, is an unusual cause of hypokalemia; it is observed most frequently in residents of the Southern United States among whom, it may be a traditional practice. The ingested clay (a kind of pica) binds K$^+$ and prevents its absorption in the gastrointestinal tract. Clay eaters who become hypokalemic usually have overall poor dietary K$^+$ intake that is additive to the binding effects of the clay.

230 Explain the mechanism of hypokalemia in patients with anorexia nervosa.

❏ Anorexia nervosa is an important clinical entity characterized by a psychological disturbance associated with poor dietary intake and self-induced vomiting. Multiple nutritional deficits, which can be life-threatening, develop in these patients.

231 Describe the causes of hypokalemia due to increased gastrointestinal loss of K^+.

❏ Substantial K^+ losses are observed with protracted vomiting, laxative abuse, diarrhea of any cause, ureterosigmoidostomy, obstructed or long ileal loop, and villous adenoma of the colon.

232 What is the mean $[K^+]$ in each of the various gastrointestinal fluids?

❏ The $[K^+]$ in pancreatic fluid, bile, and small bowel secretions is similar to that in plasma water (~5.0 meq/liter). A higher concentration, however, is found in gastric fluid (~15 meq/liter), in colonic fluid, and in diarrheal stools (~30 meq/liter).

233 Describe the mechanism for the hypokalemia observed with protracted vomiting.

❏ Protracted vomiting leads to hypokalemia that is largely caused by increased renal K^+ excretion. The increased kaliuresis is due to HCO_3^- excretion consequent to HCl-depletion (metabolic alkalosis) and to secondary hyperaldosteronism resulting from ECF volume depletion. The direct loss of K^+ as a result of vomiting is relatively small, considering that $[K^+]$ in gastric juice is ~15 meq/liter (mean value).

234 Explain how K⁺ depletion develops with diarrhea.

❑ Potassium secretion by the colon closely resembles K⁺ secretion by the distal nephron. An increased Na⁺ and fluid delivery, and mineralocorticoid excess due to volume depletion, stimulate K⁺ secretion and excretion in these two epithelia. The response of the colon during diarrhea is comparable to the response of the distal nephron to loop diuretics. Additionally, diarrhea caused by an inflammatory bowel process leads to loss of K⁺-rich mucosal cells in the stool.

235 Describe the clinical setting in which hypokalemia due to laxative abuse may be observed.

❑ Laxative abuse is often denied, hindering establishment of this diagnosis. These patients perceive laxative use as the rational approach to constipation, yet a state of chronic diarrhea is established with the concomitant K⁺ depletion. When laxative abuse is suspected, stool and urine samples should be assessed for the presence of phenophtalein or related compounds. Although diarrhea is associated with metabolic acidosis, laxative abuse produces K⁺ depletion and a variable acid-base picture (normal acid-base status, metabolic acidosis, or metabolic alkalosis). The delivery of NaCl to the colon induces K⁺ secretion and excretion on the one hand, and HCO_3^- secretion and excretion on the other. While the Na⁺ delivery to the colon increases K⁺ secretion, the delivery of Cl⁻ stimulates Cl^-/HCO_3^- exchange in the colonic mucosa, inducing HCO_3^- secretion. The resulting electrolyte deficit is K⁺ depletion with hypokalemia and metabolic acidosis.

236 Describe the hypokalemia that is caused by ureterosigmoidostomy.

❑ This procedure was performed in the past for urine diversion in obstructive uropathy resulting from urogenital abnormalities. It should be recognized that net acid excretion by the kidney is greatly undermined because its two components, namely, titratable acidity and ammonium are reabsorbed (in the colon) instead of excreted. The H^+ derived from the urinary excretion of acids titrate the HCO_3^- of alkaline intestinal secretions; the urinary ammonium is also partially reabsorbed in exchange for secreted K^+. The final effect is an increased K^+ loss and H^+ retention, inducing hypokalemia and hyperchloremic metabolic acidosis.

237 Compare the development of hypokalemia as a result of ureteroileostomy (procedure currently used for urine diversion) and ureterosigmoidostomy (old procedure performed for a similar purpose).

❑ Ureteroileostomy involves the diversion of the ureters to an ileal loop that is surgically exteriorized in the anterior abdominal wall, where the urine can be collected in a plastic bag. Ureteroileostomies are currently performed instead of ureterosigmoidostomies because the new procedure decreases the risk of developing the aforementioned electrolyte disturbances. However, if the surgically implanted ileal loop is long or becomes obstructed, hyperchloremic hypokalemic metabolic acidosis is observed, as was the case with colonic diversion of the ureters.

238 Describe the development of hypokalemia in patients with villous adenoma (and/or adenocarcinoma) of the colon.

❏ Villous adenomas belong to an uncommon group (2 to 14%) of colonic tumors that are characterized by the unrelenting production of a thick secretion that is rich in mucin and K^+. Potassium loss can be substantial and lead to severe K^+ depletion. The diagnosis of villous adenoma (and/or adenocarcinoma) of the colon must be suspected in the presence of unexplained hypokalemia, even when the patient denies diarrhea (watery diarrhea is absent in this disease).

239 Describe the causes of hypokalemia caused by an inappropriately high level of renal K^+ excretion.

❏ An abnormality in the luminal factors that control the urinary K^+ excretion (electric profile of the lumen, urine flow rate, urine pH, and the urine concentration of Na^+ and Cl^-) can result in hypokalemia due to inappropriately high kaliuresis. Alternatively, an abnormality in the peritubular factors that control urinary K^+ excretion (dietary K^+ intake, acid-base status, aldosterone) can result in hypokalemia due to excessive kaliuresis.

240 Explain the alteration in the luminal factors controlling kaliuresis which can lead to hypokalemia.

❏ A more negative electric potential in the lumen, high urine flow rate, high urine concentration of Na^+, low urine concentration of Cl^-, and high urine pH, lead to increased kaliuresis and hypokalemia. Examples for each of these categories are increased distal delivery of non-reabsorbable ions in association with avid distal nephron Na^+ absorption (e.g., large ketonuria observed in diabetic ketoacidosis, osmotic diuresis due to glucosuria, or mannitol administration), increased diuresis due to salt and water excess, increased delivery of Na^+ to the distal nephron resulting from various mechanisms (i.e., loop diuretics, thiazides, and acetazolamide), decreased delivery of Cl^- to the distal nephron (severe Cl^- depletion due to vomiting or nasogastric suction), and alkaline urine pH due to HCO_3^- excretion (milk-alkali syndrome), respectively.

241 Explain the alteration in the peritubular factors, controlling kaliuresis, which can lead to hypokalemia.

❏ A low K^+ intake in association with excessive kaliuresis due to prolonged acidosis, both respiratory and metabolic, short as well as long-lasting alkalosis (respiratory and metabolic), and stimulation of the function of the Na^+, K^+-ATPase of distal nephron cells, can lead to K^+ depletion. Examples of the first category are a low K^+ intake in association with a K^+-losing nephropathy such as renal tubular acidosis (RTA) and chronic interstitial nephritis. Examples of the second category are the increased kaliuresis observed in patients with chronic respiratory acidosis and in patients with prolonged metabolic acidosis due to diarrhea. Examples of the third category (stimulation of Na^+, K^+-ATPase of distal nephron cells) are mineralocorticosteroid excess due to primary or secondary hyperaldosteronism, and licorice abuse.

242 Describe the frequency and severity of hypokalemia with diuretic therapy.

❑ Within a week from start of diuretic therapy, a mild decrease in $[K^+]_p$ occurs (0.3 to 0.6 meq/liter), and this new $[K^+]_p$ remains constant thereafter unless an intercurrent illness that decreases K^+ intake (vomiting) or increases K^+ loss (diarrhea) develops. Hypokalemia is most commonly observed with thiazides (5% of patients) than with loop diuretics (1% of patients). The decrease in $[K^+]_p$ is directly proportional to daily dosage and duration of action of the diuretic; thus, daily administration and high dosage regimens of chlorthalidone, a long-acting thiazide, are more likely to produce severe K^+ depletion and hypokalemia. The antihypertensive effect of thiazides is achieved with small dosages (6.25 to 25 mg daily) which have a small effect on K^+ balance; consequently, high dosages of thiazides in the treatment of hypertension are not warranted since they will result in K^+ depletion without better blood pressure control.

243 Is the mechanism of hypokalemia induced by diuretics fully explained by increased kaliuresis leading to K^+-depletion?

❑ No. Mild hypokalemia can be observed in patients on diuretic therapy, in spite of having normal total body K^+ stores. Redistribution of K^+ from ECF to ICF due to activation of beta-2 adrenergic receptors accounts for the hypokalemia observed in these patients. The diuretic-induced stimulation of the sympathetic nervous system can impose an additional risk in patients prone to developing cardiac arrhythmias.

244 What is the major determinant of the large kaliuresis (leading to K^+ depletion) observed with chronic administration of both moderately high and high dosages of diuretics?

❑ Secondary hyperaldosteronism is the major determinant of the large kaliuresis and sizable negative K^+ balance observed in this setting. Secondary hyperaldosteronism is the result of NaCl depletion (ECF volume contraction) caused by the administration of large dosages of diuretics. This clinical condition is typically observed in patients with severe dietary NaCl restriction (~2 grams NaCl daily). Patients with hyperaldosteronism, whether the primary or secondary form and independently of its etiology, will develop significant K^+ depletion if given long-term diuretics, especially at high dosages.

245 Describe the hypokalemia due to renal K^+ loss that develops in Bartter's syndrome.

❑ Bartter's syndrome is an uncommon disorder, generally recognized in childhood, characterized by symptomatic hypokalemia due to K^+ depletion. These patients initially exhibit skeletal muscle weakness, polyuria, and a normal blood pressure. Laboratory data reveal metabolic alkalosis and high plasma levels of renin and aldosterone. The renal K^+ wasting is hypothesized to be due to a decreased reabsorption of K^+ and Cl^- in the loop of Henle and to increased K⁺ secretion in the distal nephron secondary to endogenous mineralocorticosteroid excess. The K⁺ depletion in this disorder can be explained partly by Mg⁺⁺ depletion due to the urinary loss of this cation.

246 What other causes of hypokalemia resemble Bartter's syndrome?

❏ When the diagnosis of Bartter's syndrome is suspected, it is mandatory to rule out more common causes of hypokalemia, including undisclosed vomiting and covert use of diuretics and/or laxatives. These various forms of hypokalemia, resembling Bartter's syndrome, are most commonly observed in otherwise healthy young women. While renal K⁺ wasting is a characteristic feature of Bartter's syndrome, this feature is absent in the syndromes of secret vomiting, laxative abuse, as well as in covert diuretic users who have recently discontinued the drug. Since patients with recent undisclosed use of diuretics and those with Bartter's syndrome have increased K⁺ excretion in the presence of hypokalemia, differentiation of these two entities may require urine analysis aimed at detection of diuretics.

247 How does Mg⁺⁺ depletion lead to K⁺ depletion?

❏ Magnesium depletion with its associated hypomagnesemia can result in renal K⁺ wasting. The excessive K⁺ loss can be due to secondary hyperaldosteronism induced by Mg⁺⁺ depletion and to direct effects of Mg^{++} depletion on cellular K^+ metabolism.

248 Describe in more detail the hypokalemia observed in states of Mg^{++} depletion.

❏ It has been proposed that severe Mg^{++} depletion leads to inappropriately high kaliuresis and consequent K^+ depletion. The cause and effect relationship between these two electrolyte disturbances, however, remains to be firmly established, especially in less severe states of Mg^{++} depletion. Low $[Mg^{++}]_p$ and $[K^+]_p$ develop in states of low electrolyte intake (e.g., alcoholism and starvation), gastrointestinal fluid loss, laxative abuse, malabsorption, and urinary K^+ loss (e.g., due to diuretics, mineralocorticosteroid excess and nephrotoxic agents like cis-platinum). Whenever hypokalemia occurs in association with hypomagnesemia, supplementation of these two ions must be secured. Magnesium depletion as well as K^+ depletion are commonly found in association with metabolic alkalosis.

249 Provide a practical classification of the severity of hypokalemia, that can be used in patient management.

❏ The following is a clinically useful classification of hypokalemia:

1. *mild* hypokalemia, which includes patients with $[K^+]_p$ between 3.0 and 3.5 meq/liter and absent or equivocal ECG changes;

2. *moderate* hypokalemia, which includes patients with $[K^+]_p$ between 2.0 and 3.0 meq/liter and/or definite ECG changes indicative of hypokalemia; and

3. *severe* hypokalemia, which includes patients with $[K^+]_p$ below 2.0 meq/liter and/or severe ECG changes, and/or profound weakness/paralysis of skeletal muscle.

250 Provide a simple approach for the differential diagnosis of hypokalemia.

❑ The first step is to firmly establish the presence of true hypokalemia in the patient by ruling out a spurious form of this electrolyte abnormality. If true hypokalemia is present, the next step is to establish whether the low $[K^+]_p$ is due to: (1) abnormal distribution of body K^+ stores; (2) K^+ depletion; or (3) a combination of these two mechanisms. The aforementioned two essential steps require a complete medical history, the performance of physical examination, and collection of laboratory data.

251 What is spurious hypokalemia?

❑ The removal of K^+ from the serum by the formed elements of the blood (i.e., leukocytes), after a blood sample has been drawn, can result in spurious hypokalemia. This condition can occur in patients with leukemia in whom K^+ uptake by the white blood cells can result in hypokalemia; this process develops at room temperature if the cells are not immediately separated from the serum, following blood drawing. It must be recognized that a high leukocyte count is associated with either spurious hypokalemia or spurious hyperkalemia.

252 How commonly is hypokalemia observed in hospitalized patients?

❑ Hypokalemia is observed in up to 50% of patients admitted to a general hospital for the care of acute diseases. Most of these patients, however, have mild hypokalemia with few or no signs or symptoms. Only one out of one thousand (1:1,000) hospitalized patients develops severe hypokalemia, with $[K^+]_p$ lower than 2 meq/liter.

253 What is mild hypokalemia observed in patients with an acute illness called? Describe its mechanism.

❑ The most commonly observed cause of hypokalemia in clinical practice is "stress hypokalemia". This form of hypokalemia results from a shift of K^+ from ECF to ICF, due to activation of beta-2 adrenergic receptors. Activation of the sympathetic nervous system leading to increased plasma catecholamine levels is found in many disease states, including exacerbations of bronchial asthma, chronic obstructive pulmonary disease (COPD), angina pectoris, myocardial infarction, gastrointestinal bleeding, febrile syndromes, and in many other processes.

254 Do the signs and symptoms of hypokalemia develop at a similar $[K^+]_p$, when this electrolyte disorder is due to redistribution of K^+, instead of its depletion?

❏ No. The effects of hypokalemia are most notable when this electrolyte disturbance is due to K^+ depletion. Since hypokalemia due to K^+ redistribution is most commonly associated with high plasma levels of catecholamines and/or insulin, it is difficult to establish with certainty the responsibility of hypokalemia for the observed manifestations.

255 Explain how to determine whether hypokalemia is due to redistribution of K^+ or, alternatively, to K^+ depletion.

❏ The patient's history and physical examination will provide the basic elements to establish whether hypokalemia is mostly caused by K^+ depletion or not. Evidence of adequate K^+ intake (i.e., K^+-rich diet), lack of gastrointestinal fluid loss (e.g., vomiting, diarrhea, suction of gastrointestinal fluids, presence of a gastrointestinal fistula), and lack of urinary K^+ loss (e.g., diuretics, cis-platinum, metabolic alkalosis, glucosuria), will favor the diagnosis of K^+ redistribution. In addition to ruling out a state of K^+ depletion, a positive history for clinical settings in which K^+ redistribution occurs must be established. A number of drugs including beta-2 adrenergic agonists (albuterol or similar agents administered orally or as inhalants), xanthines (e.g., theophylline, caffeine), toluene sniffing, Ba^{++} poisoning (ingestion of soluble barium salts, barium chloride burns), chloroquine intoxication, folate or B_{12} therapy of megaloblastic anemias, and

HCO_3^- infusion, can result in hypokalemia due to K^+ redistribution. A family history and/or a personal history of periodic episodes of paresis or paralysis, compatible with the diagnosis of hypokalemic periodic paralysis, can lead to the diagnosis of this unusual but important form of hypokalemia. Activation of the sympathetic nervous system by endogenous mechanisms is responsible for the so-called "stress hypokalemia" which is observed in the course of myocardial infarction, exacerbation of bronchial asthma, and alcohol withdrawal syndromes (i.e., delirium tremens). The role of physical examination in establishing the etiology of hypokalemia rests on the assessment of ECF volume status, skeletal muscle mass, and blood pressure (presence of normotension or hypertension).

256 Besides the medical history and physical examination, what other elements can help to establish the cause of hypokalemia?

❏ Once the diagnosis of hypokalemia due to K^+ depletion has been formulated based on the medical history, determination of the electrolyte composition of plasma and urine is the next major step to establish the cause of hypokalemia. Demonstration of high serum TCO_2 (> 32 mmol/liter) in association with hypochloremia helps to establish the diagnosis of an associated metabolic alkalosis or of chronic respiratory acidosis. Alternatively, low serum TCO_2 (< 22 mmol/liter) will help to establish the presence of metabolic acidosis or chronic respiratory alkalosis. Evaluation of electrolyte excretion requires a 24-hour urine collection. The urinary excretion of K^+ of < 30 meq/day is compatible with K^+ depletion due to extrarenal mechanisms; if K^+ excretion is > 30 meq/day in the presence of hypokalemia, excessive urine K^+ losses must play a role in the development of hypokalemia. The urine excretion of Cl^- can help in the differential diagnosis of hypokalemia due to Cl^- sensitive metabolic alkalosis (low urinary $[Cl^-]$ is observed in gastric alkalosis) or Cl^--resistant metabolic alkalosis (high urinary $[Cl^-]$ is observed in states of mineralocorticosteroid excess).

257 Does the presence of hypokalemia in association with a urinary K^+ excretion > 30 meq/day indicate renal K^+ wasting?

❏ No. "Stress hypokalemia", which is the most common cause of hypokalemia in clinical practice, occurs in association with a urinary K^+ excretion > 30 meq/day. The urinary K^+ excretion is not reduced in stress hypokalemia since total body K^+ stores are within normal limits. In contrast, the presence of hypokalemia due to total body K^+ depletion, associated with urinary K^+ excretion of more than 30 meq/day, is indicative of renal K^+ wasting, since K^+ depletion should signal the kidney to conserve K^+.

258 Describe the renal adaptation to a diminished total body K^+ content.

❏ A decreased total body K^+ secondary to a low K^+ intake or to extrarenal K^+ loss triggers renal K^+ conservation. Within 2 to 3 days from the initiation of a K^+-depleted state, urinary K^+ excretion decreases to levels lower than 30 meq/day.

The decreased urinary K⁺ secretion by the distal nephron, as well as activation of K⁺ reabsorption by the tubules, contribute to this response. Details of renal adaptation to K⁺ depletion are explained in subsequent answers.

259 Explain, succinctly, the determinants of the reduced kaliuresis observed in K⁺-depleted states in which extrarenal mechanisms cause the electrolyte derangement.

❏ While in normal individuals the fractional excretion of K⁺ (FE_{K^+}) is ~10%, K⁺ depletion sharply reduces the FE_{K^+} to levels as low as 1%. Potassium secretion by the distal nephron is drastically reduced as a result of a decreased $[K^+]_i$ in the distal tubule and in the collecting duct (due to the state of overall K⁺ depletion), as well as to diminished secretion and lower plasma levels of aldosterone (hypokalemia acts directly on the adrenal gland suppressing aldosterone release). The distal nephron in K⁺ depletion reverses its "normal" role of K⁺ secretion to become a K⁺ reabsorbing epithelium. Potassium reabsorption by the collecting duct results from an increased activity of the H⁺, K⁺-ATPase, which is a luminal membrane pump located in intercalated cells; this pump is responsible for K⁺ reabsorption in exchange for H⁺ secretion. The renal H⁺, K⁺-ATPase shares similarities with the H⁺, K⁺-ATPase of gastric mucosa. It must be noted that K⁺ secretion in the distal nephron occurs in principal cells, while K⁺ reabsorption in this segment occurs in intercalated cells.

260 Is there a clinical pattern common to all patients who develop hypokalemia due to K^+ depletion?

☐ Yes. Patients exhibiting hypokalemia due to K^+ depletion usually have two or more mechanisms that are responsible for the development of K^+ deficit. Since the mechanisms of K^+ conservation are very effective given the biological importance of K^+, a disruption of the mechanisms of K^+ homeostasis must occur at multiple sites for K^+ depletion to develop. For example, the use of diuretics (i.e., thiazides) results in clinically significant K^+ depletion when oral K^+ intake is low or when extrarenal losses of K^+ (i.e., laxatives, diarrhea) are concomitantly present.

261 Describe diagnostic possibilities in a patient with hypokalemia induced by K^+ depletion in whom urinary K^+ excretion is low.

☐ The demonstration of a low urinary K^+ excretion in a patient with K^+ depletion indicates that extrarenal mechanisms of K^+ wasting must be responsible for the disturbed K^+ homeostasis. In many instances, a low K^+ intake is a major cofactor in the development of K^+ depletion. Gastrointestinal K^+ loss is unquestionably the most commonly observed source of extrarenal K^+ depletion in clinical practice. Diarrhea, laxative abuse, and gastrointestinal fistulae are included in this category. Less frequently, skin loss of K^+ can play a significant role in K^+ depletion. Discontinuation of diuretics after prolonged use results in K^+ depletion and low urine K^+ excretion, since renal conservation of K^+ is stimulated by the K^+-depleted state.

262 What is the role of measuring blood pH in the diagnostic approach of patients with K^+ depletion due to extrarenal mechanisms of K^+ wasting?

❏ A routine determination of blood pH to establish the acid-base diagnosis with precision (i.e., metabolic acidosis, normal acid-base status, metabolic alkalosis) is not indicated. It is the patient's history that will provide information regarding the cause of the K^+-depleted state and not the precise evaluation of the acid-base status. Changes in the serum total CO_2 (TCO_2), assessed with the routine evaluation of serum electrolytes, in addition to the patient's history, will usually provide the necessary data to establish the presumptive acid-base diagnosis. A decreased serum TCO_2 (consistent with metabolic acidosis) is found in patients with gastrointestinal fistulae and diarrhea. On the other hand, an increased serum TCO_2 (consistent with metabolic alkalosis) can be seen with large skin fluid losses leading to K^+ and NaCl depletion causing secondary hyperaldosteronism. Laxative abuse is associated with normal acid-base status, metabolic acidosis, or even with metabolic alkalosis. Furthermore, the acid-base pattern observed with each of the various extrarenal mechanisms of K^+ depletion is most variable, so that precise characterization of the acid-base status with blood pH measurement fails to offer significant insight in the diagnostic approach to hypokalemia in this patient population.

263 Offer a diagnostic approach for a patient with hypokalemia caused by K^+ depletion in whom high urine K^+ excretion is found.

❏ Examination of serum electrolytes will provide a critical element in the differential diagnosis, namely whether metabolic acidosis (low serum TCO_2), metabolic alkalosis (high serum TCO_2) or normal acid-base status (serum TCO_2 within normal limits) is present. The differential diagnosis of hypokalemia associated with each of these acid-base disturbances is described in the answers formulated below.

264 What is the value of measuring blood pH in the diagnostic approach to patients with K^+ depletion due to renal mechanisms of K^+ wasting?

❏ The routine determination of blood pH to establish the acid-base diagnosis with precision (i.e., metabolic acidosis, normal acid-base status, or metabolic alkalosis) is not indicated. Changes in the serum TCO_2, assessed with the evaluation of serum electrolytes, will usually provide the necessary data to establish the presumptive acid-base diagnosis. A decreased serum TCO_2 is consistent with the diagnosis of metabolic acidosis, while an increased serum TCO_2 is consistent with metabolic alkalosis. Normal levels of serum TCO_2 are compatible with a normal acid-base status. Occasionally, measurement of blood pH may be necessary when the data obtained from the clinical history, serum and urine electrolytes, fail to establish a presumptive diagnosis.

265 What is the differential diagnosis in a patient with hypokalemia due to renal K$^+$ wasting, exhibiting a low serum TCO$_2$ (metabolic acidosis)?

❏ The clinical history will provide the most valuable elements for the differential diagnosis including: (1) toxin exposure (e.g., methanol, ethylene glycol, ethanol); (2) drugs (e.g. acetazolamide); (3) personal or family history of renal disease (i.e., RTA, proximal and distal); (4) personal or family history of diabetes mellitus (i.e., diabetic ketoacidosis); and (5) previous abdominal surgery (i.e., ureterosigmoidostomy). Calculation of the plasma anion gap from the measured serum electrolytes will help to further establish the mechanism of the HCO$_3^-$ deficit that accompanies the hypokalemia. The plasma anion gap can be high with toxin exposure and diabetic ketoacidosis; the plasma anion gap can be within normal limits in RTA, carbonic anhydrase inhibitors (e.g., acetazolamide), and ureterosigmoidostomy.

266 What is the differential diagnosis in a patient with hypokalemia due to renal K$^+$ wasting, exhibiting a normal serum TCO$_2$ (normal acid-base status)?

❏ The clinical history will provide the most valuable elements for the differential diagnosis including: (1) drugs (e.g., diuretics, fludrocortisone, Na$^+$ citrate or NaHCO$_3$, high-dose penicillin, aminoglycosides, cis-platinum); (2) recovery phase of acute tubular necrosis; (3) post-obstructive diuresis; (4) interstitial nephritis (e.g., allergic, toxic, or due to cell infiltration as in lymphomas and leukemias); and (5) states of Mg^{++} depletion.

267 What is the differential diagnosis in a patient with hypokalemia due to renal K⁺ wasting, exhibiting a high serum TCO_2 (metabolic alkalosis)?

❏ Once more the clinical history in association with physical examination will help to establish the cause of hypokalemia. We must inquire about the following factors: (1) drugs (diuretics, corticosteroids); (2) gastrointestinal disease (e.g., vomiting, nasogastric drainage, chloride-losing diarrhea); (3) previous personal history of K⁺-wasting (e.g., Bartter's syndrome); (4) presence of hypertension (e.g., primary hyperaldosteronism, congenital adrenal hyperplasia, Cushing's syndrome, Liddle's syndrome, licorice abuse); and (5) states of Mg^{++} depletion. Urine Cl⁻ levels are higher than 20 meq/liter in all the conditions listed except in patients with prior diuretic use and in cases with gastrointestinal fluid loss.

268 Should all hypokalemic patients receive K⁺ supplementation?

❏ No. Potassium supplementation given to patients with hypokalemia due to K⁺ redistribution imposes, in most instances, a serious risk of hyperkalemia, which is the most dreaded complication of K⁺ replacement therapy. On the other hand, hypokalemia caused by K⁺ depletion generally requires K⁺ supplementation. In cases of severe hypokalemia (< 2 meq/liter), whether it is the result of redistribution or to K⁺ depletion, or when it is due to a combination of these two mechanisms, K⁺ supplementation is mandatory.

269 Do all patients who are at risk of developing hypokalemia require prophylactic measures to prevent K^+ depletion?

☐ No. Preventive measures (prophylactic management) are only indicated when the risk/benefit ratio is such that benefits clearly outweigh risks. Few conditions have been recognized which deserve prophylactic K^+ replacement; these conditions include: (1) chronic use of diuretics in patients with heart disease; (2) patients on digitalis preparations with a concomitant K^+-losing condition; and (3) patients at risk of hepatic encephalopathy.

270 Elaborate further on the instances when K^+ replacement is indicated.

☐ Potassium replacement is indicated in the presence of: (1) $[K^+]_p < 3.0$ meq/liter (moderate and severe hypokalemia) with and without symptoms/associated diseases; (2) $[K^+]_p$ 3.0 to 3.5 meq/liter (mild hypokalemia) in patients with manifestations of hypokalemia, obvious depletion of body K^+ stores, or an associated disease state (heart disease) for which even mild hypokalemia can be life threatening; and (3) significant K^+ depletion and/or excessive ongoing K^+ loss (renal and/or extrarenal), or translocation into the cells (e.g., insulin effects in therapy of diabetic ketoacidosis).

271 Describe a commonly observed condition in which hypokalemia is present and K⁺ replacement is not indicated.

❑ Patients with mild asymptomatic hypokalemia ($[K^+]_p$ 3.0 to 3.5 meq/liter) due to K^+ redistribution (i.e., stress hypokalemia), who are not receiving digitalis therapy and are free of heart disease or severe liver disease, should not receive any form of K^+-replacement therapy. Thus, K^+ supplementation, K^+-sparing diuretics, or angiotensin-converting enzyme inhibitors should not be prescribed.

272 Describe the criteria for K^+ supplementation in patients with mild hypokalemia (3.0 to 3.5 meq/liter) and absent or equivocal ECG changes.

❑ Patients with mild hypokalemia, independently of the pathogenic mechanisms, should receive K^+ supplementation if K^+ loss (renal and/or extrarenal) is excessive. This approach is simply aimed at preventing a progressive negative K^+ balance in a patient who already has a mild decrement in $[K^+]_p$. Representative examples include patients recovering from surgery who have substantial gastrointestinal fluid and K^+ loss and those receiving total parenteral nutrition aimed at promoting tissue anabolism and repair.

273 What is the importance of the overall level of renal function (glomerular filtration rate) in patients who are being considered for K^+ supplementation?

❑ It is of critical importance that adequate mechanisms for K^+ excretion are present in patients considered for K^+ supplementation. The existence of adequate renal function will generally protect the patient from the extreme hyperkalemia that can result from K^+ supplementation. Patients with significant renal insufficiency (GFR as low as 20 to 30 ml/min), especially of a chronic nature, can have substantial kaliuresis providing some protection from iatrogenic hyperkalemia. Patients with acute renal insufficiency who are already in the nonoliguric phase, are also somewhat protected from iatrogenic hyperkalemia.

274 What is the importance of daily urine output (diuresis or urine volume) in patients who are being considered for K^+ supplementation?

❑ A daily urine output of at least 1 liter must be present in the course of K^+ supplementation, to provide a basic level of renal protection against iatrogenic hyperkalemia, independently of the associated GFR. The reason for this is that urine $[K^+]$ is usually severalfold the $[K^+]_p$, unless the patient is concomitantly receiving a K^+-sparing diuretic (see effect of amiloride on the electric profile in the distal nephron).

275 What is the best overall indicator of renal K^+ excretory capacity in patients about to receive K^+ supplementation?

❏ Daily urine volume is the best indicator of renal K^+ excretory capacity. Neither BUN nor plasma creatinine values provide comparable information. Lack of diuresis results in zero kaliuresis, whether blood chemistries that indicate overall renal function (BUN and creatinine) are normal or not. Furthermore, the levels of BUN and creatinine might misrepresent the current renal function in patients who are seriously ill, because these patients' chemistry is not stable (a steady-state condition has not been reached). A dramatic example is that of a patient in the immediate post-surgical period following bilateral nephrectomy (i.e., due to trauma or tumor) in whom BUN and plasma creatinine levels are normal, yet urine output and renal K^+ excretion are zero.

276 How common and serious are the risks of K^+ replacement therapy?

❏ At least one out of twenty patients receiving K^+ supplementation will develop adverse effects including significant hyperkalemia. Occasionally, life-threatening hyperkalemia is observed.

277 What patients are at greatest risk of developing hyperkalemia caused by K^+ replacement therapy?

❏ Patients with limited renal K^+ excretion are at great risk of developing hyperkalemia with K^+ supplementation. Included in this category are patients with renal insufficiency, those receiving angiotensin converting enzyme inhibitors (captopril, enalapril, etc.), or those receiving K^+-sparing diuretics (amiloride, triamterene, spironolactone), and elderly individuals (especially diabetics). Elderly patients have diminished GFR and less effective renin-angiotensin-aldosterone axis for their defense against hyperkalemia during K^+ supplementation.

278 What patients are at greatest risk of developing clinical manifestations of hypokalemia?

❏ Since the most significant effects of hypokalemia are on the heart and on the skeletal muscle, patients with diseases involving these two systems are at the greatest risk of symptomatic hypokalemia. Hence, patients with a history of atrial and ventricular arrhythmias, those receiving digitalis medication, those with cardiac enlargement and/or ischemic heart disease as well as patients with neuromuscular diseases, such as familial periodic hypokalemic paralysis and myasthenia gravis, require close monitoring of $[K^+]_p$ if a state of negative K^+ balance develops (i.e., due to diuretics, diarrhea). Potassium repletion aimed at normalizing K^+ stores and increased K^+ intake, to counterbalance the excessive ongoing K^+ loss, must be implemented.

279 Which patients, excluding those with cardiac and neuromuscular diseases, are at significant risk of developing clinical manifestations of hypokalemia?

❏ Potassium depletion can result in hepatic encephalopathy/coma and increased glucose intolerance in patients with severe liver disease and type-2 diabetes mellitus, respectively. Consequently, K^+ depletion must be avoided and K^+ repletion should be provided if hypokalemia has been documented in patients with these disorders.

280 Which patients are at greatest risk of hypokalemia and significant K^+ depletion due to prolonged use of diuretics?

❏ The use of diuretics for the treatment of edematous states (i.e., congestive heart failure, cirrhosis of the liver, and nephrotic syndrome) produces substantial urine K^+ loss; conversely, the use of diuretics in nonedematous states (e.g., hypertension) is generally associated with limited kaliuresis. Consequently, a formal plan for K^+ supplementation must be performed when long term diuretics are used for the treatment of the former group of patients.

281 Describe the overall strategy for the prevention and management of hypokalemia due to K^+ depletion.

❏ Prevention and management of a K^+ depletion state can be accomplished by either increasing K^+ intake or by decreasing K^+ loss. Potassium supplementation by the oral route, should be used to increase K^+ intake whenever possible. Excessive K^+ loss, by renal and extrarenal mechanisms, can be controlled by specific therapeutic interventions. The administration of angiotensin converting enzyme inhibitors and of K^+-sparing diuretics helps to diminish kaliuresis. A variety of antidiarrheal agents can help to diminish an excessive K^+ excretion in the stool or by an intestinal fistula.

282 Describe the commonly used preparations for oral K^+ supplementation.

❑ Several K^+ salts such as KCl, K_2HPO_4, and $KHCO_3$ (or HCO_3^- precursors such as citrate, acetate, or gluconate) are available for oral administration (liquid, powder, tablet, capsule). The strength of these preparations is as follows: each tablet/capsule (slow K, micro-K) has 8 to 10 meq; each packet of powder (K-lor, Kay Ciel, Klyte/Cl) has 15 to 25 meq; the liquid forms (KCl, Kay Ciel, Kaon) have 1 to 3 meq/ml.

283 What is the role of the accompanying anion of a K^+ salt in the correction of K^+ depletion?

❑ The anion that accompanies the K^+ salt provided in the treatment of K^+ depletion plays a role in the fate of the administered cation. Studies performed in K^+-depleted (K^+-free diet + DOCA + thiazides) animals that were nephrectomized prior to K^+ repletion with either $KHCO_3$ or KCl showed differential effects between the two anions as follows: (1) $KHCO_3$ induced a higher cellular K^+ uptake and lower steady state $[K^+]_p$; and (2) the unequal distribution of Cl^- and HCO_3^- between the extracellular and intracellular compartments was responsible for the greater cellular K^+ uptake with $KHCO_3$ infusion.

284 Describe conditions in which a particular anion must accompany the K⁺ administered in K⁺ replacement therapy.

❏ The major determinant for using a particular anion of administered K⁺ salt is the simultaneous presence of an acid-base disturbance. Potassium repletion in patients with metabolic acidosis (i.e., RTA, diarrhea) should be performed with a K⁺ salt of HCO_3^- or a HCO_3^- precursor (acetate, citrate, gluconate). Potassium repletion in patients with metabolic alkalosis (i.e., vomiting, nasogastric suction) should be performed with KCl in order to correct K⁺ and Cl⁻ depletion simultaneously.

285 Describe a specific condition where the use of potassium phosphate in patients with K⁺ depletion is indicated.

❏ Potassium phosphate (K_2HPO_4) administration is particularly useful in patients undergoing total parenteral nutrition, in whom cell anabolism and tissue repair are the major therapeutic goals. In this instance, K_2HPO_4 provides the two ions required for an efficient cellular K⁺ uptake ($HPO_4^=$ and proteins are the major intracellular anions that balance the positive charge of K⁺).

286 Give general guidelines to help decide whether parenteral administration of K⁺ should be used.

❏ The parenteral administration of K⁺ should be considered when the oral route for K⁺ replacement is either not possible or when immediate action is mandatory to correct a life-threatening condition that responds to K⁺ therapy. Examples for the mandatory use of intravenous (i.v.) K⁺ salts include patients with digitalis-induced arrhythmias or severe skeletal muscle paresis or paralysis. It must be recognized that even when immediate effects of the administered K⁺ salt are desired, an oral preparation, if available, can be used, and its effects will become evident by the time the K⁺ salt for i.v. administration is ready for infusion; whether the full dose of i.v. K⁺ salt initially considered for administration is given or not will depend on the results of a close follow up of the clinical and/or laboratory data.

287 Compare the risks of K⁺ administration by oral and parenteral routes with respect to a safe rate of K⁺ repletion and total amount of this ion given.

❑ While 50 meq of K⁺ can be safely given as a single bolus dose by oral route, a similar dose administered intravenously requires a constant infusion of five hours duration. The oral route forces administered K⁺ to go through the splanchnic circulation in which uptake by the hepatocytes and other cells prevents a large rapid entry of K⁺ to the systemic circulation. Furthermore, the oral route of K⁺ administration triggers insulin release and activates the sympathetic nervous system, thereby increasing internal disposal of the K⁺ load. Potassium administration by the oral route promotes a smoother repletion of K⁺ stores. The protective mechanisms against iatrogenic hyperkalemia during K⁺ supplementation described above for the oral route are not available with the intravenous (i.v.) administration of K⁺. Consequently, i.v. administration of K⁺ should generally not exceed 10 meq/hour. If the rate of i.v. K⁺ administration exceeds 10 meq/hour because of a life-threatening hypokalemic crisis, continuous ECG monitoring is mandatory. An exception to this rule is the K⁺ replacement therapy in the recovery from diabetic ketoacidosis.

288 What is the role of K⁺-rich foods in the prevention and management of K⁺ depletion?

❏ It is always advisable to prescribe the use of K⁺-rich foods as the initial step in the management of K⁺ depletion, prior to considering the use of oral K⁺ pharmaceutical preparations. Among the K⁺-rich foods are fruit and fruit juices (1 cup of orange juice and a medium-size banana have identical K⁺ content, 15 meq each), which have a K⁺ content equivalent to two tablets/capsules of K⁺ salts for each portion of fruit/fruit juice.

289 How is oral K⁺ supplementation usually prescribed?

❏ The usual dosage of K⁺ supplementation is 40 to 120 meq/day and should be taken with or after meals to diminish gastrointestinal side effects.

290 Is the development of hyperkalemia during i.v. K⁺ replacement indicative of normalization of body K⁺ stores?

❏ No. Severe symptomatic hyperkalemia can develop in patients receiving i.v. K⁺ supplementation despite a persistent severe K⁺ depletion. Thus repletion of K⁺ stores demands that K⁺ supplementation be given at a rate that does not exceed the speed of cellular K^+ uptake. Plasma $[K^+]$ and ECG should be monitored to correct K^+ depletion safely when i.v. K^+ therapy is instituted.

291 Describe a standard prescription for i.v. K^+ supplementation.

❑ Since available ampules of K^+ salts contain 2 meq of K^+ per ml, i.v K^+ supplementation requires dilution of this concentrated K^+ salt solution. In most conditions, 20 to 40 meq of K^+ (2 to 4 ampules) are added to each liter of saline or dextrose-containing solution. Since dextrose solutions stimulate insulin secretion, a given dose of K^+ salt will produce a smaller rise in $[K^+]_p$ when this cation is infused in a glucose-containing solution instead of a saline infusion. If the aim of therapy is to restore the depressed $[K^+]_p$ very promptly, it is advisable to dilute the K^+ ampules in NaCl-containing solutions. On the other hand, when correction of the overall K^+ deficit is the main goal of therapy, K^+ salts diluted in glucose-containing solutions are the first choice.

292 Describe the prevention and/or management of K^+ depletion by other measures, besides K^+ supplementation.

❑ Effective prevention and/or management of K^+ depletion can be accomplished with the use of K^+-sparing diuretics (spironolactone, amiloride, triamterene), as well as with angiotensin converting enzyme inhibitors. Since the major risk in the prevention and management of K^+ depletion is the development of hyperkalemia, it is unwise to superimpose two or more modes of therapy for K^+ depletion. Thus, the physician must choose the best strategy to be used in a particular patient among the various alternatives available for the prevention and/or management of K^+ depletion.

293 Compare the instances when each of the various options available to the physician for the prevention and/or management of K^+ depletion is indicated.

❑ When immediate correction (acute management/therapy) of hypokalemia is desired, the exclusive therapeutic option is K^+ supplementation. Neither the administration of K^+-sparing diuretics or angiotensin converting enzyme inhibitors have an immediate effect on $[K^+]_p$. By contrast, long term correction (chronic management/therapy) of hypokalemia can be achieved effectively and safely with the use of one of the three following options:

1. K^+ supplementation (oral, parenteral);

2. K^+-sparing diuretics (spironolactone, triamterene, amiloride); and

3. angiotensin converting enzyme inhibitors (captopril, enalapril, lisinopril).

294 Explain the relative potency/efficacy of the three options previously described in the prevention and/or long term management of K^+-depletion.

❏ The potency/efficacy of the three options, namely K^+ supplementation, K^+-sparing diuretics, and angiotensin converting enzyme inhibitors in the prevention and/or management of K^+ depletion, is of comparable value. The selection of a particular form of K^+ replacement is based on multiple factors including the patient's tolerance to any of these options (e.g., gastrointestinal intolerance to oral K^+ intake), adverse effects of medication (e.g., K^+-sparing diuretics are unsafe in diabetics, especially elderly patients), the patient's hemodynamic condition (e.g., angiotensin converting enzyme inhibitors are not indicated in patients with normal or low blood pressure), and the physician's previous experience/expertise.

295 Is there any condition in which the simultaneous use of two treatment modalities for hypokalemia is justified?

❏ Yes. Patients with persistent and clinically significant hypokalemia that fails to respond to any of the three treatment modalities used as monotherapy require a combination of two treatment modalities for K^+ repletion. It should be recognized that this particular situation is the exception and K^+ depletion in most patients is effectively managed with a single modality of K^+ repletion.

296 Describe contraindications for the use of K⁺-sparing diuretics.

❑ The major contraindication for the use of K⁺-sparing diuretics is the presence of hyperkalemia and/or renal insufficiency resulting in plasma creatinine levels > 1.5 mg/dl. Additional contraindications include the concomitant use of other drugs that interfere with renal K⁺ excretion (e.g., angiotensin converting enzyme inhibitors) and its use in elderly diabetics. Since diabetics are at increased risk of developing hyperkalemia because of insulin deficit/resistance, diabetes-induced renal insufficiency, and hyporeninemic hypoaldosteronism, these agents should be avoided in these patients, if possible.

297 Describe the use of spironolactone for the prevention and treatment of K⁺ depletion.

❑ Spironolactone inhibits Na⁺ reabsorption and K⁺ secretion in the distal nephron as a result of competitive antagonism with the action of aldosterone on this nephron segment. Its use is indicated in the syndromes of mineralocorticoid excess (primary and secondary), and in patients with hepatic cirrhosis in whom K⁺ depletion can elicit hepatic encephalopathy. The usual dosage is 50 to 100 mg/day. The major adverse effects of spironolactone are hyperkalemia, hirsutism in women, and gynecomastia in men.

298 Describe the use of amiloride in the prevention and management of K^+ depletion.

❏ Amiloride inhibits the conductance of Na^+ channels in the apical membrane of the distal nephron. The resulting inhibition of Na^+ transport prevents the development of the lumen-negative urinary space, thus reducing the electric force that pulls K^+ into the lumen. The effects of amiloride are not dependent on the presence of aldosterone. This K^+-sparing diuretic produces a generally mild natriuresis. The usual dosage is 5 to 20 mg/day.

299 Describe the use of triamterene in the prevention and management of K^+ depletion.

❏ Triamterene inhibits Na^+ reabsorption and K^+ secretion in the distal nephron, in a manner comparable to that of amiloride. This action is the consequence of direct effect on the renal tubule; it is not directly related to aldosterone antagonism or the secretion of this hormone. The usual dosage is 50 to 150 mg/day. Commercially available preparations contain triamterene in combination with hydrochlorothiazide (i.e., maxzide, diazide). The use of triamterene, amiloride, and spironolactone (group of K^+-sparing diuretics) is contraindicated when a risk of developing hyperkalemia is present.

300 What is the management of K^+ deficit in patients with diabetic ketoacidosis (DKA)?

❏ The typical patient with DKA has a K^+ deficit of ~5 to 7 meq/kg body wt at the time of admission, however the initial $[K^+]_p$ might be normal or elevated. The following practical guidelines might be used regarding the replacement of K^+:

1. To prevent hypokalemia, potassium chloride (KCl) should not be added to the initial 2 liters of i.v. infusion given to a patient with DKA, since

   a. on admission, urinary output is unknown, and

   b. initial $[K^+]_p$ is most frequently normal or high.

   Thereafter, hypokalemia tends to develop, requiring the administration of K^+ salts. The reasons for a decreased $[K^+]_p$ during therapy of DKA are multifactorial: (1) insulin-mediated cellular uptake of K^+; (2) dilution due to volume repletion by i.v. fluids; (3) correction of metabolic acidosis; and (4) urinary K^+ losses.

2. Potassium chloride (KCl) should be added to the third liter of i.v. infusion and subsequently if

   a. urinary output is adequate (it should be at least 30 to 60 ml/hour), and

   b. $[K^+]_p$ < 5 meq/liter.

   The rate of i.v. K^+ supplementation should be:

   30 to 40 meq/hour, if $[K^+]_p$ is lower than 4 meq/liter, or

   10 to 20 meq/hour, if $[K^+]_p$ is higher than 4 meq/liter.

301 How much K^+ should be added to each liter of intravenous (i.v.) solution in the therapy of DKA?

❏ The concentration of K^+ in i.v. infusions should be:

20 meq/liter when i.v. supplementation is started and subsequently, if $[K^+]_p$ is lower than 4 meq/liter and i.v. infusion rate is one liter/hour or higher, or

40 meq/liter (maximum), if $[K^+]_p$ is lower than 4 meq/liter and i.v. infusion rate is < 1 liter/hour.

302 How is the administration of K^+ monitored in a patient with DKA?

❏ Monitoring of K^+ supplementation in a patient with DKA is accomplished by obtaining measurements of $[K^+]_p$ every two to four hours, during the initial 12 to 24 hours of therapy, and by running an ECG every 30 to 60 min, during the initial four to six hours (i.e., only lead II).

303 Are there any conditions in which K^+ might be added to the initial fluid therapy in DKA?

❏ An exception to the previous rules of K^+ supplementation that mandate the addition of K^+ to the initial 2 liters of i.v. fluids must be considered if

1. the initial $[K^+]_p$ is lower than 4 meq/liter; and

2. adequate diuresis is secured.

304 Describe the management of hypokalemia in Bartter's syndrome.

❏ Since the management of the altered K^+ homeostasis associated with Bartter's syndrome is complex, multiple strategies have been proposed. The administration of nonsteroidal anti-inflammatory agents (e.g., indomethacin) inhibits the synthesis of renal prostaglandins, and decreases plasma renin activity as well as aldosterone levels; consequently, hypoaldosteronism diminishes renal K^+ excretion. Potassium-sparing diuretics (i.e., triamterene, amiloride, and spironolactone) can be useful given alone or concomitantly with other therapeutic measures. Angiotensin II converting enzyme inhibitors (e.g., captopril, enalapril, lisinopril) have been used effectively in some patients. Beta-adrenergic blocking agents (e.g., propranolol) can be of help in the management of hypokalemia by impairing the renal (i.e., depression of the renin-angiotensin-aldosterone system) and the extrarenal (i.e., depression of cellular K^+ entry) K^+ disposal. The use of orally administered K^+ supplements is a popular strategy for the management of hypokalemia in patients with Bartter's syndrome. None of the above-mentioned strategies has been clearly shown to be superior in comparison with the others.

305 Describe the management of primary or familial hypokalemic paralysis.

❏ The management of this disorder involves prophylaxis or long term therapy and treatment of acute attacks of paralysis. Acetazolamide administration, 250 to 750 mg/day, is effective in the long term therapy as a result of: (1) induction of a mild and well tolerated hyperchloremic metabolic acidosis (acetazolamide inhibits HCO_3^- reabsorption in the renal proximal tubule); (2) pH-induced effects on internal K^+ balance; and (3) direct effect of this drug on K^+ flux in cell membranes and on muscle excitability. Beta-adrenergic blockers (oral or as inhalants) are also of value in the treatment of this condition. The management of acute attacks is explained in the answer to a subsequent question.

306 Describe the treatment of Ba^{++} intoxication.

❏ Since oral ingestion of soluble barium salts is responsible for the syndrome of barium poisoning, the treatment of this condition consists of: (1) oral administration of sulfate salts (5 to 10 grams of Mg^{++} or Na^+ sulfate) to render unabsorbed soluble barium into insoluble barium sulfate; (2) gastric lavage; (3) promotion of renal excretion of Ba^{++} with i.v. saline infusion and furosemide; and (4) parenteral administration of K^+.

307 Describe the management of secondary hypokalemic paralysis.

❏ Since all forms of secondary hypokalemic paralysis are the result of severe K^+ depletion and the associated hypokalemia, their management is replenishment of K^+ stores. The therapeutic modality chosen to repair and sustain adequate K^+ levels (i.e., K^+ supplementation, K^+-sparing diuretics, angiotensin converting enzyme inhibitors) depends upon the etiology of K⁺ deficit, the patient's tolerance, and the physician's expertise with each of these therapeutic regimens.

308 Describe the most salient characteristics of thyrotoxic skeletal muscle paralysis.

❏ As opposed to familial/primary periodic paralysis, patients with the thyrotoxic form experience their initial episode of paralysis when they are adults (almost always older than 20 years). It is most commonly observed in patients with specific HLA haplotypes that are characteristically observed in individuals of Asian descent. Most commonly, patients develop hypokalemia during paralytic crisis, although this is not a constant feature. In a fashion that mimics the familial form, the episodes of paresis/paralysis are commonly triggered by a high carbohydrate meal, cold exposure, and rest immediately after exercise. The management of thyrotoxic skeletal muscle paralysis includes the administration of propranolol (a beta-adrenergic antagonist that can also help to control loss of muscle strength observed in familial hypokalemic paralysis) and antithyroid agents.

309 Let us review a clinical history. A 44-year-old woman was brought to the hospital with progressive weakness of four weeks duration. Previous history was significant for dry mucous membranes for which a workup yielded a diagnosis of Sjögren's syndrome. Physical examination disclosed a major decrease in muscle strength of all extremities accompanied by reduced deep tendon reflexes. Other abnormalities were not found. Laboratory data were as follows:

Hematocrit	48 %
plasma creatinine	1.3 mg/dl
BUN	20 mg/dl

Serum Electrolytes:

$[Na^+]_p$	136 meq/liter
$[K^+]_p$	1.7 meq/liter
$[Cl^-]_p$	116 meq/liter
total CO_2 (T_{CO_2})	10 mmol/liter
anion gap*	10 meq/liter

* $[Na^+]_p - ([Cl^-]_p + [T_{CO_2}]_p)$

Arterial blood gases:

pH	6.96 ($[H^+]$ ~110 nmol/liter)
P_{CO_2}	45 mmHg
P_{O_2}	70 mmHg

Urine:

specific gravity	1.014
pH	6.8
protein	1+

What is the most likely pathogenesis of the clinical syndrome including profound hypokalemia?

❏ The patient's symptoms and findings on physical examination are best explained by hypokalemia due to K^+ depletion. Her previous history is negative for a poor dietary K^+ intake, gastrointestinal fluid loss, excessive sweating, and medication that can result in high urinary K^+ excretion (e.g., diuretics). Since the patient did not have previous episodes of muscle weakness/paralysis and laboratory data revealed additional distinct abnormalities, the diagnosis of familial hypokalemic periodic paralysis is ruled out. Consequently, increased renal K^+ excretion is the most likely cause of hypokalemia, and the pathogenesis of the clinical syndrome is best explained by a primary renal disease.

310 What is the specific diagnosis of this patient (described in the previous question), who initially exhibited progressive skeletal muscle weakness?

❏ The skeletal muscle dysfunction is due to K^+ depletion. The overall renal function (glomerular filtration rate –GFR– assessed by the patient's BUN and plasma creatinine levels) was not significantly depressed, however severe acidemia due to a mixed metabolic acidosis and respiratory acidosis was present. The greatest alteration in the acid-base composition was a decreased $[HCO_3^-]_p$ that was associated with a normal plasma anion gap (normal plasma unmeasured anions), indicating that the decreased HCO_3^- stores were not the result of titration by an excessive retention of anions (high anion gap metabolic acidosis due to renal failure, ketoacidosis, etc.). The patient's hyperchloremic metabolic acidosis was not accompanied by the expected hypocapnia (secondary respiratory response to a primary metabolic acid-base disturbance) since the P_{CO_2} was in the upper limit of normal; consequently, the acid-base diagnosis is that of a mixed acid-base disturbance in which hyperchloremic metabolic acidosis is accompanied by respiratory acidosis. The respiratory depression is best explained by dysfunction of respiratory muscles caused by hypokalemia and K^+ depletion. The specific diagnosis that defines the overall clinical syndrome, including the hypokalemia and the severe acid-base disturbance, is a form of renal tubular acidosis.

311 What type of renal tubular acidosis (RTA) best explains the symptoms, signs, and laboratory data of the patient described in the previous two questions?

❏ Distal or type-1 RTA is the most likely diagnosis based on the: (1) previous history of Sjögren's syndrome; (2) severe metabolic acidosis with normal plasma anion gap, also known as hyperchloremic metabolic acidosis; (3) overall renal function (assessed by GFR) relatively unimpaired; (4) abnormally high urine pH (patient's urine pH was 6.8) for the severity of metabolic acidosis since blood pH was below 7.0; if renal acidification mechanisms were intact, urine pH should have been at its lowest value (< 5.0); and (5) fact that the renal acidification defect was located in the distal nephron, since this segment is responsible for maximally lowering the urine pH when plasma acidity is severely increased. In proximal or type-2

RTA, the distal acidifying mechanism is normal resulting in a very low urine pH when $[HCO_3^-]_p$ and blood pH are significantly decreased; and 6) hypokalemia due to renal K^+ wasting, which is a typical finding in distal or type-1 (classic) RTA.

312 Describe the treatment of hypokalemia due to renal tubular acidosis (RTA).

❑ Type-1 RTA (classic or distal) and type-2 RTA (proximal) can result in hypokalemia due to excessive urinary K^+ loss. While hypokalemia and increased kaliuresis is mild in untreated patients with proximal RTA, the distal variety (type-1 RTA) commonly leads to severe K^+ depletion in untreated patients. The administration of large amounts of HCO_3^- required for the treatment of proximal RTA induces increased kaliuresis leading to K^+ depletion, unless K^+ supplementation is provided in significant amounts. In contrast with proximal RTA, the requirement of alkali and K^+ supplementation in distal RTA is modest; a daily dosage of 40 to 80 meq/day of K^+ and alkali are usually sufficient to maintain K^+ and acid-base balance in distal RTA, while larger amounts of both substances must be used in the treatment of proximal RTA. Potassium HCO_3^- (or bicarbonate-forming salts of K^+) is a widely used preparation for the management of both proximal and distal RTA.

313 Describe a state of tissue hypoexcitability generated by the simultaneous presence of two electrolyte disturbances.

❑ A low $[K^+]_p$ (hypokalemia) in association with a high $[Ca^{++}]_p$ (hypercalcemia with high ionized calcium level) can result in life-threatening tissue hypoexcitability.

314 Describe a clinical condition characterized by decreased tissue excitability due to the simultaneous presence of two electrolyte disorders.

❑ Advanced neoplasias with bone metastases characteristically result in hypokalemia (poor K^+ intake) and hypercalcemia (Ca^{++} released from bones); the former hyperpolarizes the resting potential and the latter depolarizes the threshold potential. Consequently, a profound depression of neuromuscular excitability develops due to widening of the difference between resting and threshold potentials of the cell membrane.

315 Let us review another clinical history. An 82-year-old woman was found lying on the floor of her home by neighbors who brought her to the emergency room. She was in a stuporous state and had stool and urine on her clothing. Previous history was unavailable, but neither physician's prescriptions nor actual medications were found in her belongings or house cabinets. Physical examination showed sunken eyes and decreased skin turgor, indicative of severe volume depletion (reduction of ECF volume). Blood pressure was 72/10 mmHg; pulse, 102/min supine; it was considered unwise to move the patient to a standing position for examination of the hemodynamic response. Neck examination revealed flat jugular veins; respirations were 42/min, and neurological examination revealed no focal deficit. The patient's body wt was 42 kg. Laboratory studies showed:

plasma creatinine   2.6 mg/dl
BUN   41 mg/dl

Serum electrolytes:

[Na^+]   132 meq/liter
[K^+]   2.2 meq/liter
[Cl^-]   113 meq/liter
total CO_2 (TCO_2)   7 mmol/liter
anion gap*   12 meq/liter
    *$[Na^+]_p - ([Cl^-]_p + [TCO_2]_p)$

Arterial blood gases:

pH   7.26 ([H^+] ~55 nmol/liter)
PCO_2 16 mmHg
PO_2 76 mmHg

Provide a general assessment of this patient's condition and the need for immediate therapy.

❑ This elderly woman is critically ill considering her hemodynamic condition (significant depletion of body fluids, known as volume depletion, as well as hypotension), her altered mental status, and deranged laboratory values (hypokalemia, metabolic acidosis, and prerenal azotemia). Her medical management should include repletion of lost body fluids, which should help correct hypotension and improve perfusion of vital organs (i.e., brain, heart). The patient's hypokalemia and metabolic acidosis require immediate attention, thus i.v. fluid replacement therapy must include K^+ and alkali supplementation.

316 What is the most likely mechanism of hypokalemia in this elderly, obtunded woman?

❑ The severe hypokalemia observed in this patient is most likely due to depletion of K^+ stores rather than to abnormal K^+ distribution in the setting of normal body K^+ stores. Her negative history in the use of medication with the potential to induce hypokalemia (e.g., insulin administration, beta-adrenergic agonists, xanthines), the severity of hypokalemia, associated volume depletion and metabolic acidosis, strongly point toward K^+ depletion as the cause of hypokalemia.

317 What is the specific cause of hypokalemia observed in this critically ill patient?

❑ Large intestinal K^+ loss in association with insufficient K^+ intake best explain the patient's K^+ depletion. Although diuretic therapy is the most commonly observed cause of severe hypokalemia, especially in the elderly, evidence of prior diuretic use was lacking (neither physician's prescriptions nor these drugs were found on the patient's night table or in her kitchen/bathroom cabinets). In addition, the abnormal acid-base status found in this patient (metabolic acidosis) is not to be found in patients taking the most widely used diuretics (thiazides and loop diuretics produce metabolic alkalosis). Further evidence of gastrointestinal K^+ loss was the observation of fecal staining of her clothes. The laboratory data, revealing hyperchloremic metabolic acidosis accompanying hypokalemia, is most consistent with fecal losses of alkali and K^+. Although other presumptive diagnoses including RTA are possible, the severity of metabolic acidosis in the absence of a previous history of renal disease in this elderly woman, makes these alternative diagnoses most unlikely.

318 Would you predict the K^+ deficit in this elderly, volume depleted woman to be mild or severe? Estimate total body K^+ deficit.

❑ The patient's total body K^+ deficit is severe, and amounts to a few hundred meq. If we were to apply the slope (0.3 meq/liter $[K^+]_p$ decreases for every 100 meq K^+ loss) of the relationship between the change in $[K^+]_p$ and that of body K^+ content, the K^+ deficit would be:

$$\Delta[K^+]_p = \text{normal } [K^+]_p - \text{actual } [K^+]_p$$
$$= 4.0 - 2.2$$
$$= 1.8 \text{ meq/liter}$$

$$K^+ \text{ deficit} = \frac{1.8 \text{ meq/liter}}{0.3 \text{ meq/liter}/100 \text{ meq}}$$
$$= 600 \text{ meq}$$

It must be recognized that the slope 0.3 meq/liter/100 meq overestimates K^+ loss in this elderly woman with small muscle mass. Yet, acidemia in our patient promotes a smaller decrement in $[K^+]_p$ in response to K^+ depletion because of pH-mediated movement of K^+ from ICF to ECF. Consequently, these opposing effects would counterbalance each other.

319 What should be the fluid therapy and its route of administration in this patient?

❑ This elderly woman who manifested severe water and electrolyte depletion cannot be repleted by the oral route either effectively (because of associated diarrhea) or safely (because of increased risk of aspiration pneumonia due to altered mental status). Thus, the route of choice is the intravenous (i.v.) administration of fluids. The simultaneous correction of water and electrolyte abnormalities is properly handled by the administration of a 5% solution of dextrose in water with 0.45% NaCl (D5W ½ normal saline) plus one ampule of $NaHCO_3$ (45 meq) and 30 meq KCl/liter of i.v. solution. This fluid regimen will help correct the decreased levels of K^+, HCO_3^- and Na^+, and increased levels of Cl^-, BUN, and creatinine. The initial rate of i.v. infusion should be ~200 ml/hr in order to promptly restore her depleted intravascular volume; once hypotension is corrected, it is wise to decrease the i.v. infusion rate to ~100 ml/hour.

320 Let us review another case. A 72-year-old woman with a history of diabetes mellitus and arteriosclerotic cardiovascular disease was brought to the emergency room in a coma. Her medication included digoxin and insulin for the treatment of heart failure and diabetes mellitus, respectively. Admission laboratory values were as follows:

plasma creatinine	1.7 mg/dl
BUN	36 mg/dl
serum glucose	820 mg/dl
serum ketones	4+ in a 1:2 dilution

Serum electrolytes:

$[Na^+]$	128 meq/liter
$[K^+]$	4.0 meq/liter
$[Cl^-]$	94 meq/liter
total CO_2 (TCO_2)	5 mmol/liter

Arterial blood gases:

pH	7.02 ($[H^+]$ ~95 nmol/liter)
PCO_2	20 mmHg
PO_2	82 mmHg

What is your assessment of this patient's body K^+ stores?

❏ The patient's K^+ stores are significantly decreased. The normal $[K^+]_p$ can be misinterpreted as indicative of adequate K^+ stores; however, hyperglycemia and acidemia in this diabetic patient characteristically produce K^+ depletion associated with normal $[K^+]_p$. The severe hyperglycemia causes massive glucosuria which is associated with large renal K^+ losses that lead to severe K^+ depletion. The abnormal internal K^+ balance caused by insulin deficit and acidemia, promotes cellular K^+ exit. Recognition of K^+ depletion in this patient is most important since insulin administration to correct the diabetic decompensation (hyperglycemia and high anion gap metabolic acidosis) induces cellular K^+ uptake; thus, profound hypokalemia can develop, unless K^+ supplementation is also provided.

321 Let us review another clinical case. A 52-year-old woman known to have liver cirrhosis, presumably resulting from heavy ethanol intake, was brought to the emergency room by her son in a confused state. The patient's son reported that she had not drunk any alcohol during the last 2 years. The patient had been taking an unknown medication during the previous few weeks that had been prescribed by her new doctor who had been consulted one month prior to admission because of increasing abdominal girth. The patient's son also reported that she had been eating "as usual", that she did not have any diarrhea, and that she had lost ~5 kg over the past 3 weeks. Vital signs were: blood pressure, 102/50 mmHg; pulse, 120/min; and respirations, 23/min. Significant physical findings included flapping tremor (asterixis) of both hands; (1+) peripheral edema, and ascites. Laboratory data on admission were as follows:

plasma creatinine	1.7 mg/dl
BUN	35 mg/dl

Serum electrolytes:

$[Na^+]$	126 meq/liter
$[K^+]$	2.7 meq/liter
$[Cl^-]$	77 meq/liter
total CO_2 (TCO_2)	33 mmol/liter

What is the most likely explanation for the confused state and flapping tremor in this patient?

☐ Hepatic encephalopathy is the most likely explanation for the development of an altered mental status and flapping tremor. Accumulation of waste products normally cleared by the liver is responsible for the neurological syndrome; a state of acute liver insufficiency has developed in this patient known to have chronic liver disease.

322 What is the most likely explanation for the development of hepatic (liver) encephalopathy in this middle aged woman?

❑ Based on the clinical history and physical examination, neither an infectious (e.g., viral hepatitis), nor a toxic (e.g., overdose of acetaminophen), or a hemodynamic (e.g., liver underperfusion due to cardiogenic shock or portal vein thrombosis) insult on this patient's liver was present. A weight loss of ~5 kg in a three-week period that coincided with a newly prescribed medication was recognized; this observation provides the clue to the etiology of hepatic encephalopathy. Deterioration of liver function is commonly observed in patients with chronic liver disease receiving aggressive diuretic therapy. The patient's large weight loss, electrolyte abnormalities (i.e., hypokalemia, hyponatremia, metabolic alkalosis) and abnormal renal function (i.e., increased levels of BUN and plasma creatinine) are best explained by intense use of diuretics; this etiology was confirmed as a result of a phone call made to the patient's family doctor who indicated that he had prescribed furosemide 80 mg twice a day. Diuretic-induced K^+ depletion is the most important determinant of the patient's hepatic encephalopathy that brought her to the emergency room.

323 Explain the mechanisms that lead to hepatic encephalopathy due to aggressive use of diuretics in patients with severe liver disease.

❏ The use of most diuretics results in increased kaliuresis and leads to K^+ depletion; K^+ deficiency, in turn, increases ammonia (NH_3) production by the kidney which is largely retained in the body fluids of patients with severe liver disease. Salt and water retention by the kidney are stimulated in patients with severe liver disease, especially after diuretic-induced volume depletion, resulting in decreased delivery of fluid to the distal nephron; it is precisely in the distal nephron where renal NH_3 is converted to NH_4^+ (ammonium), which is trapped in the urine and finally excreted. Since less Na^+ reaches the distal nephron where it promotes distal tubule acidification, a reduced trapping of NH_3 / NH_4^+ occurs, leading to diffusion of NH_3 into the renal venous blood (as opposed to the urine), entering the systemic circulation. While the normal liver rapidly clears NH_3 in the process of urea synthesis, the abnormal liver allows a build-up of NH_3 in body fluids. Hyperammonemia leads to cellular dysfunction which is largely responsible for the development of hepatic encephalopathy. In summary, diuretics given to patients with severe liver disease can result in hypokalemia, hyperammonemia, and hepatic encephalopathy/coma.

324 What is the proper management of this patient that was admitted with hepatic encephalopathy, severe electrolyte derangements, and renal insufficiency?

❏ Since abnormal nitrogen metabolism with accumulation of NH_3 and other amines is responsible for the pathogenesis of hepatic encephalopathy, the aim of therapy is to reduce their body levels. Accumulation of nitrogen products should be corrected by: (1) reducing the absorption of these compounds by the gastrointestinal tract; (2) increasing the renal excretion of products of nitrogen metabolism (urea and NH_4^+); and, (3) increasing the liver capacity for NH_3/amines uptake and subsequent metabolic disposal. Reduction of gastrointestinal absorption of nitrogen compounds involves: restriction of dietary protein intake; mechanical cleansing of the colon by enemas; oral administration of antibiotics to inhibit bacteria in the colon with urease-producing capacity (urease of microorganisms reverses hepatic conversion of NH_3 to urea so that a buildup of NH_3 occurs in the bowel leading to its subsequent absorption and NH_3 intoxication); and oral administration of lactulose, an inert sugar that acidifies the colonic content trapping NH_3 as NH_4^+, and produces diarrhea, thereby significantly increasing the NH_4^+ excretion by the gastrointestinal tract.

An increased renal excretion of nitrogen products is accomplished by correcting the overall renal function, which was depressed due to salt and water depletion and associated electrolyte abnormalities. Intravenous infusion of fluids containing NaCl and KCl will help correct prerenal azotemia, Cl^- and K^+ deficits, and metabolic alkalosis.

An improved liver function is of utmost importance for the patient's recovery, and it helps diminish plasma levels of NH_3/amines. This goal can be achieved by optimal nutrition with the administration of glucose, essential amino acids (with an overall low protein intake), K^+, and vitamins.

325 Let us review another clinical history. A 38-year-old man consulted a physician because of vomiting and diarrhea of three days duration. On physical examination he was found to have hypertension (blood pressure 150/102 mmHg). The patient had only received symptomatic medication since viral gastroenteritis had been diagnosed. Two weeks later a follow-up visit to his physician revealed that the patient was free of gastrointestinal symptoms, yet his blood pressure remained elevated. During the ten-day period prior to the follow-up visit, the patient received no medications and ingested a normal diet. To assess the new onset of hypertension (the patient had a negative history for this condition), the following laboratory data were obtained:

plasma creatinine	0.9 mg/dl
BUN	15 mg/dl

Serum electrolytes:

$[Na^+]$	145 meq/liter
$[K^+]$	2.7 meq/liter
$[Cl^-]$	94 meq/liter
total CO_2 (TCO_2)	37 mmol/liter
anion gap*	14 meq/liter

*$[Na^+]_p - ([Cl^-]_p + [TCO_2]_p)$

Is viral gastroenteritis (which brought the patient in for the initial consultation) the most likely explanation for hypokalemia and associated electrolyte disturbances observed in this patient?

☐ No. It seems reasonable, however, to first consider whether the viral gastroenteritis that led the patient to the initial medical consultation, can explain the observed electrolyte abnormalities. The patient's hypokalemia was associated with increased serum TCO_2, in the absence of respiratory dysfunction leading to chronic hypercapnia; thus, the most likely explanation for elevated serum TCO_2 is the presence of metabolic alkalosis (normal serum TCO_2 is 21 to 30 mmol/liter). Since the patient had had a normal dietary intake during the ten days immediately preceding collection of laboratory data, the possible electrolyte deficits resulting from loss of gastrointestinal fluids (and the possible development of metabolic alkalosis) should have been dissipated, so that a normal electrolyte profile should have been found. The elevated blood pressure manifested during two physician visits is an unexpected new finding for a patient with fluid loss due to viral gastroenteritis; furthermore, the observed hypertension provides a clue as to the presence of hypokalemia and other electrolyte abnormalities induced by mechanisms unrelated to the gastrointestinal fluid loss.

326 What disease entities should be considered in the differential diagnosis of this patient who had hypertension, hypokalemia, and metabolic alkalosis?

☐ The disease entities responsible for the development of hypokalemia, metabolic alkalosis, and hypertension include the syndromes of corticosteroid excess. The laboratory data will separate these patients into the four following groups, according to the plasma levels of renin and aldosterone:

1. high renin and high aldosterone,

   these patients have a primary renal process that enhances renin levels and, secondarily, stimulates aldosterone secretion by the adrenal gland. On the contrary, if the primary process of hypersecretion of aldosterone occurs in the adrenal gland, a secondary inhibition of renin secretion will occur and plasma renin levels will be low. The renal process might be hypersecretion of renin due to a tumor, hyperplasia, or renal ischemia (renal artery stenosis, accelerated hypertension, or intrarenal vascular disease);

2. low renin and high aldosterone,

   these patients have a primary adrenal process that enhances plasma aldosterone levels and suppresses plasma renin levels. This category includes primary aldosteronism (tumor of the adrenal gland), pseudoprimary aldosteronism (bilateral hyperplasia of the adrenal glands), dexamethasone suppressible hyperaldosteronism, and adrenal carcinoma;

3. low renin and low aldosterone,

   the presence of either exogenous intake of drugs or the existence of abnormal pathways in the synthesis of corticosteroids explain this category. The following processes are included: chronic administration of steroids, licorice abuse, swallowing the substances extracted from chewing tobacco, Liddle's syndrome, and a number of primary adrenal diseases (11-hydroxylase deficiency, 17-hydroxylase deficiency and adrenal carcinoma); and

4. low renin and variable aldosterone,

the presence of ACTH excess and/or Cushing syndrome are characteristic examples of this category of disorders.

Clinical Disorders

2. Hyperkalemia

327 Define hyperkalemia.

❏ Hyperkalemia is the condition where $[K^+]_p$ is above 5.0 meq/liter.

328 What is pseudohyperkalemia?

❏ Pseudohyperkalemia is the condition where $[K^+]_p$ is normal and overall K^+ homeostasis is preserved, yet the laboratory results disclose an abnormally high $[K^+]_p$. Spurious hyperkalemia, also known as pseudohyperkalemia (artifactual or false hyperkalemia), must be ruled out when the diagnosis of hyperkalemia is entertained. Pseudohyperkalemia can be induced by ischemic blood drawing, thrombocytosis (platelet count > $10^6/mm^3$), leukocytosis (WBC count > $10^5/mm^3$), in vitro hemolysis, and in a rare hereditary disorder (familial pseudohyperkalemia). This hereditary disorder does not have clinical signs and symptoms and simply expresses K^+ release from red cells when a blood sample is exposed to low temperatures in vitro.

329 Provide a broad description of the most salient effects of hyperkalemia.

❏ Hyperkalemia impairs multiple body functions, conveniently classified in the following categories: (1) cardiovascular; (2) neuromuscular; (3) renal; and (4) endocrine/metabolic.

330 What are the effects of hyperkalemia on cardiac excitability?

❑ Hyperkalemia alters cardiac excitability by depolarizing the resting MP. Hyperkalemia also increases the cell membrane permeability to K^+, shortening the duration of the action potential, since cellular exit of K^+ is responsible for returning the depolarized cell membrane to its resting state. Furthermore, pacemaker activity is depressed since the spontaneous depolarization characteristic of pacemaker cells is inhibited.

331 Describe a state of tissue hyperexcitability resulting from the simultaneous presence of changes in resting and threshold potentials that are due to two electrolyte disturbances.

❑ A high $[K^+]_p$ (hyperkalemia) in association with a low $[Ca^{++}]_p$ (hypocalcemia with low ionized calcium level) can induce life-threatening tissue hyperexcitability as a result of cardiac arrhythmias, generalized convulsions, and skeletal muscle paralysis. Hyperkalemia depolarizes the resting MP (moves the potential toward zero) while hypocalcemia hyperpolarizes (moves in a negative direction, away from zero) the threshold potential. Thus, the threshold and resting potentials are brought closer together, increasing the electric excitability of the cell.

332 Describe a clinical condition characterized by increased tissue excitability due to the additive effects of two electrolyte disorders.

❏ Renal failure produces hyperkalemia and hypocalcemia; the former depolarizes the resting MP and the latter hyperpolarizes the threshold MP, as described above. Consequently, a state of tissue hyperexcitability develops that is potentially fatal.

333   Compare the effects of hyperkalemia and hypokalemia on the electrophysiologic properties of the heart.

❏ While hyperkalemia characteristically depresses most electric properties of the heart, hypokalemia increases these properties. Thus, automaticity, duration of action potential, refractory period, and resting membrane potential, are all decreased in hyperkalemia and increased in hypokalemia. The altered electrophysiologic properties of the heart with these two electrolyte abnormalities are responsible for the development of arrhythmias.

334   Describe how hyperkalemia impairs pacemaker activity.

❏ The process of spontaneous depolarization characteristic of pacemakers results from a persistent imbalance between cellular Na^+ entry and cellular K^+ exit. The entry of a number of cations (Na^+) greater than the exit of other ones (K^+) is responsible for the spontaneous depolarization found in pacemaker cells, in the normal state. Hyperkalemia inhibits the spontaneous depolarization because it elicits a persistent high permeability to K^+ leading to increased cellular exit of this ion, therefore counterbalancing the depolarization induced by Na^+ movement into the cell. Consequently, hyperkalemia inhibits the automaticity of the cardiac pacemaker.

335 Describe the effect of hyperkalemia on the T wave of the electrocardiogram (ECG).

❑ The T wave of the electrocardiogram is the expression of ventricular repolarization. Hyperkalemia produces changes in the T wave that include narrowing its base and an increase in its amplitude ("peaked" T wave). This effect, observed when $[K^+]_p$ is above 6 meq/liter, is the result of ventricular repolarization of shorter duration.

336 Elaborate further on the cardiovascular effects of hyperkalemia.

❑ The cardiac effects of hyperkalemia are usually the dominant manifestations and most frequently precede other signs and symptoms. ECG monitoring allows recognition of the characteristic changes including peaked T waves, prolonged PR interval, depressed ST segment, shortened QT interval, widening of the QRS complex, absence of P waves, development of "sine waves", and possible asystole or ventricular fibrillation. Hyperkalemia also causes dilation of the blood vessels.

337 Describe the neuromuscular effects of hyperkalemia.

❑ Hyperkalemia produces neuromuscular dysfunction involving both skeletal and smooth (vascular, gastrointestinal, and urinary) muscle. Paresthesias and generalized skeletal muscle weakness are observed with moderate hyperkalemia. A more profound neuromuscular dysfunction that includes respiratory failure and flaccid paralysis is occasionally present in patients with severe hyperkalemia. Deep tendon hyporeflexia develops, but sensory deficit is not observed. Paresis and/or paralysis of smooth muscle produce vasodilation and decreased motility of the gastrointestinal and urinary tracts. The major manifestation of hyperkalemia that leads the patient to medical attention is occasionally neuromuscular.

338 Compare the clinical manifestations of hypokalemia and hyperkalemia with regard to skeletal muscle.

❑ Weakness, paresis, paralysis, fasciculations, and hyporeflexia/areflexia are common features in the two electrolyte abnormalities. Hypokalemia is also associated with cramps and myalgias; when K^+ depletion is severe, rhabdomyolysis can develop. By contrast, in some instances, hyperkalemia is associated with hyperreflexia and myotonia (primary/familial hyperkalemic periodic paralysis).

339 What are the electrolyte disturbances responsible for muscle weakness and those that can result in paralysis?

❑ Multiple electrolyte abnormalities (abnormally high and low levels of K^+, Na^+, Ca^{++}, and Mg^{++}, and low phosphate levels) can cause weakness. However, they are more severe with hypokalemia, hypophosphatemia, and hypercalcemia. On the other hand, few electrolyte disturbances cause paralysis, namely, hypokalemia, hyperkalemia, and hypomagnesemia.

340 Describe the renal effects of hyperkalemia.

□ Hyperkalemia alters the renal excretion of Na^+, K^+, and H^+. Sodium and K^+ excretion are increased, the latter due to increased distal nephron K^+ secretion. Hyperkalemia decreases HCO_3^- reabsorption in the proximal tubule, renal ammoniagenesis, and overall net acid excretion.

341 Describe the endocrine/metabolic effects of hyperkalemia.

□ Hyperkalemia is a potent direct stimulus of the adrenal gland to secrete aldosterone and of the pancreas to secrete insulin. Renin activity and prostaglandin synthesis are depressed, and catecholamine production is increased.

342 Provide a practical classification of the severity of hyperkalemia that can be used in patient management.

□ The following is a clinically useful classification:

1. *mild* hyperkalemia, which includes patients with $[K^+]_p$ between 5.0 and 6.0 meq/liter, and absent or equivocal electrocardiographic (ECG) changes;

2. *moderate* hyperkalemia, which includes patients with $[K^+]_p$ between 6.0 and 7.0 meq/liter, and/or

definite ECG changes in ventricular repolarization (narrow and tall T waves, also referred to as "peaked" T waves); and

3. *severe* hyperkalemia, which includes patients with $[K^+]_p$ above 7.0 meq/liter, and/or severe ECG abnormalities including atrial standstill, advanced A-V heart block, intraventricular block (QRS widening) or ventricular arrhythmias, and/or profound weakness/paralysis of skeletal muscle.

343 What are the general mechanisms leading to hyperkalemia?

❏ Hyperkalemia can be due to K^+ redistribution from ICF to ECF in the presence of normal or even decreased total body K^+ stores. Alternatively, hyperkalemia can result from an increase in total body K^+ stores in states of K^+ overload.

344 Describe the general causes of hyperkalemia due to redistribution of body K^+ stores.

❏ Hyperkalemia occurring in the presence of normal or low total body K^+ stores is indicative of abnormal mechanisms for the control of the internal K^+ distribution (abnormal internal K^+ balance). These abnormalities include disturbances in hormones (insulin, glucagon, catecholamines), the acidity of body fluids, the level of other electrolytes, the tonicity in body fluids, and drugs. Hyperkalemia due to K^+ redistribution also occurs with cell lysis as a result of leakage of K^+-rich cytosolic fluid to ECF.

345 What are the specific causes of hyperkalemia due to redistribution of body K^+ stores?

❏ The specific causes of hyperkalemia within this category include insulin deficiency, administration of beta-adrenergic blockers, acidosis, hypertonicity of ECF, and drugs such as digitalis, succinylcholine, and arginine-HCl.

346 What are the mechanisms of drug-induced hyperkalemia?

❏ Drug-induced hyperkalemia can occur as a result of a disruption of either the internal K^+ balance or of the external K^+ balance (decreasing urinary K^+ excretion).

347 How do drugs impair the internal K^+ balance and thereby pose a risk for hyperkalemia?

❏ Since 98% of the K^+ stores are within the cells, drugs that decrease cellular K^+ uptake or promote K^+ exit from cells can produce hyperkalemia. Prominent examples of these drugs are alpha-adrenergic agonists, beta-adrenergic blockers, digitalis, and muscle relaxants (succinylcholine, suxamethonium). Drugs that induce cell lysis (of tumors, red cells, and muscle tissue) can produce severe hyperkalemia. Arginine-HCl, used in the treatment of hepatic encephalopathy and of metabolic alkalosis, can also elevate $[K^+]_p$.

348 List drugs that impair the external K^+ balance and can therefore cause hyperkalemia.

❑ Drugs that diminish renal K^+ excretion include the K^+-sparing diuretics (spironolactone, amiloride, and triamterene), angiotensin II converting enzyme inhibitors (captopril, enalapril, lisinopril, etc.), nonsteroidal anti-inflammatory agents (indomethacin, ibuprofen, etc.), heparin, and cyclosporin. Any drug that causes renal failure can lead also to hyperkalemia.

349 How do digitalis preparations (digoxin, digitoxin) alter internal K^+ balance, possibly leading to hyperkalemia?

❑ Digitalis inhibits the Na^+, K^+-ATPase in all tissues including those of the heart and skeletal muscle. Consequently, cellular uptake of K^+ diminishes and $[K^+]_e$ tends to rise. However, significant hyperkalemia is only found in the presence of a large overdose of these drugs arising from accidental or intentional intake. Extreme hyperkalemia leading to death can occur.

350 Describe further the hyperkalemia that can develop in digitalis intoxication.

❑ Considering that digitalis inhibits the Na^+, K^+-pump, less K^+ is transported into the cells when toxic plasma levels of this drug are present. The diagnosis of digitalis-induced hyperkalemia requires a high index of suspicion, electrocardiographic (ECG) abnormalities suggestive of digitalis toxicity, and confirmation of toxic plasma levels of the digitalis preparation.

351 What are the characteristics of succinylcholine-induced hyperkalemia?

❑ Since succinylcholine is a muscle relaxant that depolarizes skeletal muscle, increasing K^+ permeability of its cells, a modest increment in $[K^+]_p$ occurs in most patients receiving this drug. However, severe hyperkalemia is found in some patients with neuromuscular diseases, and those with increased total body K^+ stores.

352 Explain further the hyperkalemic effects of skeletal muscle relaxants (succinylcholine, suxamethonium).

❑ Hyperkalemia, occasionally very severe, can occur with the use of skeletal muscle relaxants (depolarizing agents), and is caused by the exit of K^+ from skeletal muscle. These agents are administered most frequently in the course of general anesthesia. In most cases, the rise in $[K^+]_p$, when present, is mild and short-lasting.

353 Describe clinical settings in which hypertonicity of body fluids is responsible for the development of hyperkalemia.

❑ A hypertonic glucose infusion produces hyperkalemia in diabetics but not in normal individuals. Hypertonic i.v. infusions of other substances including NaCl, mannitol, or radiocontrast media can produce hyperkalemia in both normal individuals and diabetic patients, especially if renal K^+ excretion is compromised.

354 Explain the mechanism of chemotherapy-induced hyperkalemia.

❏ The administration of chemotherapeutic agents (e.g., cyclophosphamide, vincristine) for the treatment of various malignancies can result in rapid tumorlysis leading to cellular K^+ exit and hyperkalemia. The rise in $[K^+]_p$ is prompt but short lasting, and is most commonly seen in patients with leukemia and lymphomas, after intravenous administration of these drugs.

355 Describe the hyperkalemia that occurs in hyperkalemic periodic paralysis.

❏ Hyperkalemic periodic paralysis is an uncommon hereditary disease transmitted as an autosomal dominant trait. Hyperkalemia can occur in this syndrome with increased K^+ intake, cold weather, exercise, or at rest. The attacks of paralysis, however, are not always associated with hyperkalemia.

356 What are the clinical forms of hyperkalemic skeletal muscle paralysis?

❏ The most common clinical form of hyperkalemic paralysis is the so-called secondary hyperkalemic paralysis, which is a consequence of hyperkalemia of any cause (i.e., adrenal insufficiency, renal insufficiency, etc.) The other clinical presentation of hyperkalemic paralysis is known as primary hyperkalemic periodic paralysis. The primary form is an extremely rare familial disease in which attacks of paresis/paralysis occur in association with K^+ efflux from skeletal muscle. Thus, hyperkalemia is induced by an abnormal internal K^+ balance of unknown cause since abnormalities in insulin secretion, glucagon, catecholamines, mineralocorticosteroids, or glucocorticosteroids, are not present in this condition; furthermore, during attacks of paralysis, external K^+ balance remains unchanged or even becomes negative.

357 How persistent and severe is hyperkalemia resulting from redistribution of body K^+ stores?

❏ Hyperkalemia resulting from K^+ redistribution is usually short lasting and, if this process occurs in the presence of normal renal K^+ excretion, its severity is from mild to moderate. However, hyperkalemia due to redistribution might occasionally be severe and even lethal.

358 How frequently does a single abnormality in the internal K^+ disposal cause substantial hyperkalemia?

❏ Infrequently. The existence of several physiologic mechanisms aimed at preserving a normal $[K^+]_p$ prevents the occurrence of hyperkalemia when one of these mechanisms fails since the remaining mechanisms can compensate for the failure. Two or more alterations must be present simultaneously to consistently induce hyperkalemia. Therefore, insulin deficiency in association with acidosis and digitalis intake in association with beta-adrenergic blockers are examples of additive effects of deranged regulatory mechanisms of internal K^+ distribution that can lead to hyperkalemia.

359 Describe causes of hyperkalemia due to cellular damage.

❏ Many conditions such as intense catabolic rate, hemolysis, rhabdomyolysis, and lysis of tumor cells can cause hyperkalemia due to potassium leaking from cells to ECF. Other causes of hyperkalemia are internal bleeding and resolving hematomas.

360 Explain how hyperkalemia due to K^+ overload can develop.

❏ Potassium overload leading to hyperkalemia can occur because of an increased K^+ intake or because of decreased renal K^+ excretion. The high K^+ intake-induced hyperkalemia occurs when the adaptive increase in renal K^+ excretion is insufficient to match the larger than normal K^+ intake.

361 Elaborate on hyperkalemia caused by increased K⁺ intake.

❏ Body K⁺ excess can result from oral intake of K⁺-rich meals, salt substitutes, K⁺ supplementation, K⁺-containing pharmacologic agents, and transfusion of stored packed red blood cells or blood.

362 What are so-called salt substitutes?

❏ Salt-substitutes are K⁺ salts (KCl) that mimic the taste of NaCl. Whenever a patient is told to reduce his or her NaCl (salt) intake, these salt-substitutes are often recommended. Since these products are available over-the-counter, patients frequently use them whether they are recommended by physicians or not. A low NaCl intake reduces the ability of the kidney to excrete K⁺ to its maximum; simultaneous ingestion of salt-substitutes (K⁺ salts) can lead to hyperkalemia due to the combination of increased K⁺ intake and reduced renal K⁺ excretion.

363 Name a commonly used pharmacologic agent that contains significant amounts of K⁺.

❏ The parenteral administration of large doses of penicillin in the form of K⁺-salt (K⁺-penicillin) represents a source of substantial K⁺ intake. Potassium content of 1 million units of K⁺-penicillin is ~1.7 meq; therefore, a patient receiving 30 million units of penicillin/day will be receiving a supplementary 51 meq of K⁺ daily, as a result of this therapy.

364 Succinctly describe the various mechanisms that can induce hyperkalemia due to a decreased renal excretion of K⁺.

❏ The following processes are responsible for a decreased renal K⁺ excretion:

1. oliguria or anuria of any cause;

2. decreased GFR;

3. decreased tubular secretion of K⁺;

4. hypoaldosteronism or pseudohypoaldosteronism; and

5. drugs.

365 Describe the causes of hyperkalemia caused by inappropriately low renal K⁺ excretion.

❏ An abnormality in the luminal factors that control the urinary K⁺ excretion (electric profile of the lumen, urine pH, urine flow rate, and urine concentration of Na⁺ and Cl⁻) can produce hyperkalemia due to K⁺ retention. Alternatively, an abnormality in the peritubular factors that control the urinary K⁺ excretion (dietary K⁺ intake, acid-base status, aldosterone) can cause hyperkalemia due to K⁺ retention.

366 Explain the alteration in luminal factors that control kaliuresis which can lead to hyperkalemia.

❏ Loss of lumen negativity (luminal voltage close to 0 mV), low urine flow rate, low urine concentration of Na^+, high urine concentration of Cl^-, and an unfavorable pH for K^+ secretion lead to K^+ retention and hyperkalemia. Examples for each of these categories are K^+-sparing diuretics (amiloride, triamterene, and spironolactone), water and salt depletion, decreased delivery of Na^+ to the distal nephron resulting from the action of multiple mechanisms (e.g., increased proximal reabsorption of salt in congestive heart failure), increased delivery of Cl^- to the distal nephron (NH_4Cl administration), and an acid urinary pH, respectively.

367 Explain the alteration in peritubular factors controlling kaliuresis that can lead to hyperkalemia.

❏ A high K^+ intake, and an impairment in the function of the Na^+, K^+-ATPase of distal nephron cells, can lead to hyperkalemia. Examples of the first category include the use of salt substitutes (KCl) to enhance the flavor of a NaCl-free diet, the administration of K^+-rich pharmacologic agents (K^+-penicillin), oral K^+ supplementation (e.g., K-dur, Kaon), and K^+-rich parenteral nutrition. Examples of the second category are the syndromes of aldosterone deficit and resistance to this hormone.

368 Describe the mechanisms of hyperkalemia due to diminished renal K$^+$ excretion in the absence of generalized renal failure (normal GFR).

❏ In the absence of generalized renal failure, a diminished renal K$^+$ excretion can be due to either a defect in the renin-angiotensin-aldosterone-axis or to renal resistance to aldosterone.

369 Describe the syndromes of diminished K$^+$ excretion induced by defects in the renin-angiotensin-aldosterone-axis.

❏ The syndromes of diminished mineralocorticosteroid activity include the following groups of defects: (1) angiotensinogen deficiency; (2) renin deficiency; (3) converting enzyme deficiency; (4) defective angiotensin II receptors; and (5) adrenal insufficiency.

370 What are the causes and effects of angiotensinogen deficiency?

❏ Angiotensinogen deficiency causes hypoaldosteronism that reduces renal K$^+$ excretion. The production of angiotensinogen by hepatocytes is diminished with glucocorticosteroid deficiency (chronic adrenocortical insufficiency or Addison's disease). Advanced liver failure can also lead to hyperkalemia due to angiotensinogen deficiency.

371 Describe the causes and effects of renin deficiency.

❏ Renin deficiency generates a low plasma aldosterone level that reduces renal K^+ excretion. Renin deficiency occurs in certain physiologic states (advanced age, expansion of ECF volume), with the use of various drugs (beta-adrenergic blockers, inhibitors of prostaglandin synthesis, methyldopa), with certain toxins (lead), in some systemic diseases (diabetes mellitus), and in some renal diseases (obstructive uropathy, interstitial nephritis). A common cause of renin deficiency seen in clinical practice is the so-called type-4 RTA characterized by impaired excretion of both K^+ and H^+. Perhaps its most common form is observed in elderly diabetics.

372 In what disease states is type-4 RTA most likely to occur and what is the most common form observed?

❏ Type-4 RTA occurs in disease states characterized by aldosterone deficiency or resistance, and results in impaired K^+ and H^+ excretion. Perhaps the most common form is observed in elderly diabetics with hyporeninemic hypoaldosteronism. Other causes of type-4 RTA are obstructive uropathy, sickle cell nephropathy, and renal transplant rejection.

373 Describe the causes and effects of converting enzyme deficiency.

❏ Converting enzyme deficiency leads to hypoaldosteronism which reduces renal K^+ excretion.

The widely used antihypertensive agents, captopril, enalapril, and lisinopril, belong to this category of drugs that can increase $[K^+]_p$ levels by this mechanism.

374 Explain the defect in the renin-angiotensin-aldosterone axis which is caused by defective angiotensin II receptors.

❑ Angiotensin II (AII) interacts with receptors in the adrenal gland to stimulate aldosterone secretion. The condition known as defective AII reception is characterized by a failure of the adrenal gland to secrete aldosterone in response to stimulation by AII, leading to hyperkalemia. In this condition, ACTH (adrenocorticotrophic hormone) remains capable of stimulating aldosterone secretion.

375 Provide a succinct description of the steps involved in the renin-angiotensin-aldosterone axis.

❑ Renin is a protein, synthesized in the kidney by the juxtaglomerular apparatus, and released into the circulation, that enzymatically cleaves angiotensinogen. The renin substrate angiotensinogen is a protein synthesized in the liver. Renin action on angiotensinogen results in the formation of angiotensin I (AI) which is further cleaved by the converting enzyme (a process that occurs mostly in the lung) resulting in the production of AII. The secretion of aldosterone by the adrenal cortex (zona glomerulosa) is stimulated by AII. Aldosterone released by the adrenal gland acts on the renal tubules where it enhances the transport of Na^+, K^+, and H^+.

376 Describe the hormonal defect that leads to hyperkalemia in patients with adrenal insufficiency.

❏ Adrenal insufficiency leading to hyperkalemia can result from either aldosterone deficiency, cortisol deficiency, or a combination of these two defects. Selective aldosterone deficiency occurs with failure in production/release/action of any component of the axis including angiotensinogen, renin, AI, converting enzyme, AII, or adrenal enzyme defects. Cortisol deficiency (cortisol is secreted by the zona fasciculata of the adrenal gland) can occur after prolonged use of exogenous corticosteroids and in diseases that damage or destroy the adrenal cortex (Addison's disease, and adrenal hemorrhage or metastasis).

377 Enumerate the syndromes of diminished aldosterone activity.

❏ A diminished aldosterone activity can occur as a result of: (1) a primary defect in the adrenal synthesis of aldosterone due to a disease/defect in the adrenal cortex; this adrenal abnormality is the cause or initial event that leads to the defect in K^+ homeostasis; (2) a secondary defect in the adrenal synthesis of aldosterone due to failure in the production/release/action of the various components in the cascade leading to stimulation of the adrenal synthesis of aldosterone (i.e., renin deficiency, etc.); and (3) end-organ resistance to aldosterone, due to either drugs acting on the kidney (e.g., spironolactone), or renal disease.

378 Describe causes of a primary defect in the adrenal synthesis of aldosterone.

❏ A primary defect in aldosterone synthesis occurs with generalized adrenal failure (decreased secretion of all adrenal hormones) and with a selective defect in the production of aldosterone. Generalized adrenal failure (Addison's disease) is found in autoimmune diseases, amyloidosis, malignancies, and in chronic infection of the adrenal gland (e.g., tuberculosis). Suppression of adrenal function by exogenous corticosteroids (e.g., oral prednisone) or by acute structural damage of the adrenal gland due to infarction or hemorrhage within the gland, also result in non-selective adrenal insufficiency. A selective defect in the production of aldosterone occurs with various enzyme deficiencies and with the administration of some drugs (e.g., heparin).

379 Describe drug-induced mechanisms of hypoaldosteronism leading to hyperkalemia.

❏ Hyperkalemia induced by hypoaldosteronism can be observed with the administration of prostaglandin synthetase inhibitors, cyclosporin, and heparin. Prostaglandin synthetase inhibitors such as the nonsteroidal anti-inflammatory drugs (e.g., indomethacin) and cyclosporin inhibit renin secretion producing hyporeninemic hypoaldosteronism. Angiotensin-converting enzyme inhibitors (e.g., captopril, enalapril, etc.) decrease plasma levels of angiotensin II. Finally, heparin acts directly on the adrenal gland inhibiting aldosterone secretion (increased $[K^+]_p$ can occur with heparin administration in ~5% of hospitalized patients).

380 What is the mechanism of cyclosporin A-induced hyperkalemia?

❑ Cyclosporin A is a potent immunosuppressive agent that can induce hyperkalemia due to selective aldosterone deficiency caused by inhibition of renin secretion. In addition, this drug has the potential to cause significant nephrotoxicity, therefore leading to a more severe K^+ retention.

381 Elaborate further on the mechanism of cyclosporin-induced hyperkalemia.

❑ Cyclosporin can produce hyperkalemia as a result of a diminished renal K^+ excretion. The cyclosporin-induced K^+ retention is explained by depression of the renin-angiotensin-aldosterone system and/or direct renal injury resulting in diminished renal function (tubular injury and decreased glomerular filtration rate).

382 Describe how nonsteroidal anti-inflammatory drugs (NSAID) can produce hyperkalemia.

❑ NSAID can diminish renal K^+ excretion, leading to hyperkalemia. The mechanism of K^+ retention by NSAID includes the inhibitory effect of these drugs on prostaglandins of renal origin, which in turn diminishes glomerular filtration rate (GFR) and renin release (which leads to decreased aldosterone secretion). NSAID can also have a direct nephrotoxic effect leading to renal failure, thereby causing additional renal K^+ retention.

383 How does heparin increase $[K^+]_p$?

❑ Large and small doses of heparin administered by either subcutaneous or intravenous routes induce K^+ retention and increase renal Na^+ excretion. These effects result from heparin-induced inhibition of aldosterone secretion by the adrenal gland. The severity of the hyperkalemia is usually moderate.

384 What are the causes of end-organ resistance to aldosterone, leading to hyperkalemia?

❑ The syndrome of hyperkalemia due to end-organ resistance to aldosterone is known as pseudohypoaldosteronism. This entity, characterized by a renal resistance to aldosterone effects, can develop as a result of drug administration and/or renal diseases. Hyperkalemia caused by spironolactone, triamterene, and amiloride, collectively known as K^+-sparing diuretics, exemplify drug-induced pseudohypoaldosteronism. Renal diseases that primarily damage the renal tubules and spare the glomeruli (or compromise GFR only mildly), collectively known as tubulointerstitial renal diseases, elicit this hyperkalemic syndrome. Renal diseases included in this category are sickle-cell disease, amyloidosis, Sjögren's syndrome, systemic lupus erythematosus, and obstructive uropathy. Hyperkalemia develops frequently in renal transplant recipients and aldosterone resistance of the allograft is partly responsible for it. In addition, there are patients with pseudohypoaldosteronism in whom the cause is unknown (idiopathic pseudohypoaldosteronism).

385 Explain the mechanism of hyperkalemia caused by K^+-sparing diuretics.

❑ Spironolactone, amiloride, and triamterene inhibit electrolyte transport in the distal nephron, the renal segment responsible for the bulk of urinary K^+ excretion. The primary action of these drugs is inhibition of Na^+ transport which secondarily diminishes the electric potential that drives K^+ (a positively charged ion) from the distal tubule cells to the urinary lumen. While amiloride and triamterene act directly on the renal tubules, spironolactone acts indirectly by inhibiting the effects of aldosterone on the distal nephron.

386 What is idiopathic pseudohypoaldosteronism?

❑ This disease, an extremely rare entity, includes two forms known as type I and type II (Gordon's syndrome). While type I is a disease of infants, type II is observed in adults. Infants with type I disease have a generalized lack of response to aldosterone in the excretory (renal tubules), alimentary (salivary glands, colon), and integumentary (sweat glands) systems. Some characteristics of type I disease are impaired growth and development, as well as renal salt wasting. The adult type II patient, characteristically has hypertension and impaired K^+ excretion caused by increased Cl^- reabsorption in the distal nephron (i.e., chloride shunt).

387 Does decreased renal function always cause K⁺-retention leading to hyperkalemia?

❏ No. In the presence of renal insufficiency, potassium balance is maintained within normal limits until the glomerular filtration rate (GFR) decreases to less than 25% of normal value. The ability of the kidney with decreased GFR to maintain K⁺ balance depends on the development of compensatory mechanisms that are collectively known as K⁺ adaptation.

388 Describe the process of K⁺ adaptation in renal insufficiency.

❏ The increased efficiency of the remaining nephrons (functional unit of the kidney consisting of the glomerulus and its tubules) to excrete K⁺ develops rapidly (within hours) after a reduction in renal tissue. This response involves the cortical, medullary, and papillary collecting tubule. The "principal cells" of these segments increase the number of K⁺ pumps (Na⁺, K⁺-ATPase), allowing the cells to increase the transport rate of these ions as well as to increase urine K⁺ excretion per nephron.

389 Compare the K⁺ excretion rate in normal individuals and those with renal failure.

❏ When $[K^+]_p$ remains constant (whether low, normal, or high), overall K⁺ excretion (renal and extrarenal) is identical to K⁺ intake. The individual is therefore described as being in a steady-state condition. Consequently, patients with renal insufficiency can have low, normal, or even high total rates of K^+ excretion depending on the concomitant K^+ intake. The excretion of K^+ in renal failure is not necessarily different from that in the normal state since both groups must have identical K^+ excretion and intake in steady-state conditions. However, a high K^+ intake can lead to progressive hyperkalemia in patients with renal insufficiency while in normal individuals it is well tolerated.

390 Should treatment be initiated for mild (less severe degree) hyperkalemia?

❑ Yes. Hyperkalemia of any severity requires prompt therapeutic intervention because of the potential risk of a life-threatening state caused by cardiac arrest (asystole or ventricular fibrillation) and/or respiratory arrest (skeletal muscle paralysis).

391 What therapeutic measures should be initiated for all forms (mild, moderate, or severe) of hyperkalemia?

❑ The following measures should be started immediately upon recognition of hyperkalemia:

1. discontinuation of all sources of K^+ intake, oral and parenteral;

2. interruption of the administration of drugs that can impair K^+ homeostasis leading to hyperkalemia, including beta-adrenergic blockers, prostaglandin inhibitors, and K^+-sparing diuretics;

3. removal of the toxic substances that have played a role in the development of hyperkalemia (e.g., digitalis drugs); and

4. promotion of a negative K^+ balance by increasing K^+ excretion through renal and extrarenal routes.

392 Describe the dietary management of patients predisposed to the development of, or already affected with, abnormal body K^+ stores.

❏ A K^+-rich diet should be prescribed to patients at risk of developing or already suffering K^+ depletion and hypokalemia. In a comparable manner, a K^+-poor diet is indicated for patients predisposed to, or already exhibiting body K^+ excess and/or hyperkalemia. In order to adjust the dietary K^+ intake to the patient's need for this cation, it is mandatory to instruct the patient in the subject of K^+ content of various foods.

393 Provide examples of K^+-free foods that are of special value in patients predisposed to hyperkalemia.

❏ Examples of K^+-free foods include sugar, olive oil, butter, honey, and marmalades (e.g., blueberries, cranberries).

394 Provide examples of low-K^+ foods that are of special value in patients predisposed to hyperkalemia.

❑ Examples of low-K^+ foods (K^+-poor) include: (1) grain products such as bread, rice, and pastas; (2) eggs and milk products (cheese); (3) certain fruits such as apples, pears, and tangerines; and (4) certain vegetables such as lettuce and onions. These foods have less than 10 meq of K^+ per cup.

395 Provide examples of K^+-rich foods, the intake of which should be limited in patients predisposed to hyperkalemia and encouraged in those predisposed to hypokalemia.

❑ Examples of K^+-rich foods include: (1) red and white meat as well as fish; (2) most legumes (i.e., beans, peas); (3) most fruits and their juices (i.e., bananas, oranges, cantaloupes, grapes, nectarines, pineapples, watermelons, tomatoes, papayas, prunes, cherries, lemons, peaches, plums, strawberries, avocados); (4) potatoes (regular and sweet), mushrooms, carrots, corn, artichokes, cabbage; and (5) chocolate. These foods have 10 to 30 meq of K^+ per cup.

396 Provide examples of very K^+-rich foods.

❑ These foods are of special value for use in K^+-depleted patients but are strictly forbidden for patients predisposed to hyperkalemia. Examples of very K^+-rich foods include: (1) dried fruits (e.g., raisins, dates, apricots, prunes); (2) nuts (e.g., peanuts, pecans, almonds, walnuts); and (3) spinach, and cocoa powder. These foods have more than 30 meq of K^+ per cup.

397 Does bowel transit have any major influence on the hyperkalemia that is associated with end-stage renal failure?

❏ Yes. Potassium adaptation in patients with end-stage renal failure stimulates K^+ secretion by the colon, which in this setting is the major route of K^+ excretion. This route can excrete 30 meq/day or more if daily bowel movements occur. The daily use of stool softeners and the occasional use of laxatives and enemas avoid constipation and optimize K^+ excretion by the colon in patients with renal failure.

398 Describe the therapeutic measures for hyperkalemia aimed at increasing renal excretion of K^+.

❏ In order to promote increased kaliuresis, all patients, except those with end-stage renal failure or physical signs of fluid overload/pulmonary edema must have substantial NaCl intake via oral or parenteral routes. In addition, the renal excretion of K^+ can be increased by the administration of a single diuretic agent (furosemide, thiazides, acetazolamide), or a combination of these agents. Diuretics should not be administered to patients with end-stage renal failure since a meaningful kaliuretic response is not expected.

399 How do you promote a negative K^+ balance for the treatment of hyperkalemia in patients with end-stage renal failure?

❑ A negative external K^+ balance is achieved in patients with end-stage renal failure by: (1) the utilization of cation exchange resins (such as Kayexalate, a polystyrene sulfonate) that promote the excretion of K^+ in the stools; and, (2) dialysis (hemodialysis and peritoneal dialysis).

400 What is the most commonly observed contributing factor in the development of hyperkalemia in clinical practice?

❑ A low NaCl intake is the single most commonly observed contributing factor in the development of hyperkalemia. Removing the salt restriction promotes increased kaliuresis which might partially or fully correct the hyperkalemia.

401 What is the proper management with respect to dietary NaCl intake of patients with a salt-retaining disease (hypertension, congestive heart failure, cirrhosis of the liver, or nephrotic syndrome) who have an elevated $[K^+]_p$ or even a high normal (i.e., 5.0 meq/liter) $[K^+]_p$?

❑ Proper management of simultaneous retention of Na^+ and K^+ in these patients is effectively achieved by avoiding severe restriction of dietary NaCl intake while concomitantly administering diuretics (e.g., furosemide, thiazides). This recommendation with respect to the dietary NaCl intake should be instituted once a large ECF volume excess is no longer present. A moderate dietary NaCl intake of ~4 grams/day will not result in ECF volume expansion if an increased urine excretion of NaCl is achieved with the use of diuretics. This strategy of enhancing salt (NaCl) intake and excretion secures adequate kaliuresis (delivery of salt to the distal nephron is reduced with a salt-restricted diet and predisposes to hyperkalemia).

402 What therapeutic measures should be considered in patients who have high-normal $[K^+]_p$ (i.e., 5.0 meq/liter) and a disease that predisposes to hyperkalemia?

❑ Several measures should be undertaken at once in these patients. Severe restriction of dietary NaCl intake should be avoided since it impairs renal K^+ excretion; dietary NaCl intake should be at least 4 grams/day. Restriction of dietary K^+ must be enforced. Medications that impair the renal excretion of K^+, including converting enzyme inhibitors (i.e., captopril, enalapril, lisinopril) and K^+-sparing diuretics (i.e., triamterene, amiloride, and spironolactone) should be discontinued. Metabolic acidosis, if present, should be treated with alkali therapy.

403 Describe the therapy of type-4 RTA.

❑ Since hyperkalemia is the major threat to life in patients with type-4 RTA, the main focus of attention should be placed on correcting this electrolyte abnormality. That is the reason why dietary K^+ restriction, diuretics (furosemide, thiazides), and K^+ binding resins are so valuable. The intake of

NaCl should be encouraged, since the availability of Na^+ salts in the collecting tubules is a major determinant of renal K^+ excretion. The administration of fludrocortisone (Florinef) in daily doses of 0.1 to 0.3 mg helps in the correction of hyperkalemia and acidosis (enhances distal acidification); yet, the associated volume expansion may induce hypertension or increase its severity. Alkali therapy (1 to 2 meq/kg/day) is useful to ensure correction of hyperkalemia and acidosis.

404 Describe the management of hyporeninemic hypoaldosteronism and Gordon's syndrome, entities characterized by impaired kaliuresis leading to hyperkalemia in the presence of relatively well-preserved GFR.

❏ The syndrome of hyporeninemic hypoaldosteronism, commonly seen in diabetics, is effectively managed with adequate salt intake plus furosemide, and/or the administration of fludrocortisone (a synthetic mineralocorticosteroid agent). Treatment of hyperkalemia in Gordon's syndrome is accomplished by the administration of thiazide diuretics.

405 Describe general strategies for the treatment of severe hyperkalemia.

❏ The three following strategies must be considered whenever severe hyperkalemia is present: (1) counterbalance the effect of hyperkalemia on the excitability of myocardial and skeletal muscle with the administration of drugs. This modality is not aimed at reducing the increased $[K^+]_p$; (2) modify internal K^+ balance, promoting the translocation of K^+ from ECF to ICF. This modality will not alter the total body K^+ stores; and (3) modify external K^+ balance, inducing a net K^+ loss from the body.

406 Does proper treatment of severe hyperkalemia entail the sequential use of the various therapeutic modalities previously described?

❏ No. Severe hyperkalemia is a life-threatening electrolyte abnormality that mandates the concerted action of all available therapeutic maneuvers. Thus, in the presence of severe hyperkalemia, several therapeutic options must be implemented at once.

407 Describe the treatment of hyperkalemia with agents that ameliorate the effects of hyperkalemia on myocardial and skeletal muscle excitability.

❏ Administration of Ca^{++} salts (chloride or gluconate) will diminish tissue excitability by widening the difference between resting and threshold potential. When Ca^{++} salts are provided, the resting MP will remain depolarized by the hyperkalemia while the threshold MP will be depolarized by the Ca^{++}, enlarging the difference between resting and threshold MP, so its value returns toward normal. Calcium gluconate (20 ml of a 10% solution) can be infused intravenously (i.v.) over a 10 min period. The i.v. administration of Ca^{++} salts is definitely indicated when $[K^+]_p$ reaches 7.0 meq/liter and/or when significant ECG abnormalities (absence of P wave, prolongation of QRS complexes, etc) are present. The effects of Ca^{++} infusion are short-lasting with peak effect noted about five min after infusion.

408 Describe the therapy of hyperkalemia aimed at inducing the translocation of K^+ from ECF to ICF.

❏ The most important pharmacologic agent that promotes cellular K^+ entry is insulin. Since insulin promotes tissue uptake of K^+ and glucose, the latter must be infused to prevent hypoglycemia. Considering that hyperglycemia of endogenous and exogenous origin can result in hyperkalemia (especially in diabetics), caution should be exercised as to the rate of glucose infusion. Consequently, a situation that mimics an euglycemic insulin clamp (providing exogenous glucose in adequate amounts to maintain normal plasma glucose level) must be instituted. A less important strategy that can translocate K^+ to ICF is $NaHCO_3$ infusion (described in a subsequent answer).

409 Describe in detail the use of insulin and of glucose in the treatment of severe hyperkalemia.

❏ Insulin can be effectively administered by means of an intravenous (i.v.) bolus of 5 U regular insulin followed by a continuous i.v. insulin infusion; the latter might consist of 10 units of regular insulin dissolved in 500 ml of a solution of 20% glucose in water, infused at a rate of 100 ml/hour. Plasma glucose and $[K^+]_p$ should be closely monitored with measurements taken every 30 min to evaluate the results of therapy and to prevent major changes in plasma glucose. The stated rate of insulin/glucose mixture can be modified to reach therapeutic goals.

410 What is the effect of $NaHCO_3$ infusion on $[K^+]_p$?

❏ The administration of sodium bicarbonate (alkali) is expected to move K^+ into cells inducing hypokalemia by raising extracellular pH. However, the effect of HCO_3^- infusion on $[K^+]_p$ is inconsistent. Furthermore, the mild hypokalemia observed with this alkali infusion has been shown to occur not from the transfer of K^+ to ICF but rather from the dilution of extracellular K^+ due to expansion of the ECF volume (the currently used ampules of $NaHCO_3$ are one normal solutions having 1 equivalent/liter; thus, their osmolality is about six times that of body fluids, therefore promoting fluid transfer from ICF to ECF).

411 What is the role of $NaHCO_3$ infusion in the treatment of hyperkalemia?

❏ Hyperkalemia associated with severe acidemia due to metabolic acidosis (primary reduction in $[HCO_3^-]_p$) should be treated with $NaHCO_3$ infusion unless the patient has significant ECF volume expansion and high risk of pulmonary edema. However, $NaHCO_3$ infusion is generally an unreliable tool for the control of severe hyperkalemia, especially in patients with renal failure in whom bicarbonaturia and the associated kaliuresis will not develop. Thus, HCO_3^- administration should not be considered the primary modality but only a supplementary one in the therapy of severe hyperkalemia. A combined protocol that includes insulin plus glucose administration (primary treatment modality) and alkali infusion (supplemental treatment modality) can be as follows: one to two ampules (44 meq/ampule) of $NaHCO_3$ and 10 units of regular insulin added to a solution of 20% glucose in water for infusion at 100 ml/hour.

412 What is the role of hypertonic NaCl infusion in the treatment of hyperkalemia?

❏ Hypertonic NaCl infusion can be used for the treatment of hyperkalemia in hyponatremic patients (those with low serum sodium concentration). Although hypertonicity of ECF, especially with mannitol, i.v. radiocontrast media, and hyperglycemia (i.e., diabetics) can induce or aggravate hyperkalemia, hypertonic NaCl infusion can help to reverse the abnormal electric profile resulting from hyperkalemia. The possible salutary effects of hypertonic NaCl in the treatment of hyperkalemia rest on: (1) the induction of a more alkaline cell pH by enhancing Na^+/H^+ counter exchange (Na^+ enters the cell, H^+ exits the cell); (2) an increased difference between threshold and resting membrane potential, achieved when the increased Na^+ gradient between $[Na^+]_p$ and $[Na^+]_i$ moves Na^+ into the cells and promotes cellular Ca^{++} entry (the effect is comparable to that of Ca^{++} infusion); (3) the repolarization of the resting potential by cellular entry of Cl^- which favors cellular re-entry of K^+.

413 Describe another pharmacologic agent for the treatment of severe hyperkalemia that promotes the transfer of K^+ from ECF to ICF.

❏ Cellular K^+ uptake is increased with the administration of selective beta-2 adrenergic agonists. Albuterol, a member of this family of drugs, can be administered (20 mg dosage) as an aerosol for inhalation. This treatment modality, although effective, must be considered only as supplementary.

414 Describe the time frame of the therapeutic response to insulin and glucose, $NaHCO_3$, and albuterol, in the management of severe hyperkalemia.

❏ The therapeutic response to insulin and glucose infusion and to albuterol is prompt, with a significant hypokalemic effect observed within 60 min. By contrast, the effect of $NaHCO_3$ infusion on $[K^+]_p$, if any, requires a few hours.

415 What is the importance of inducing a net K^+ loss (negative external K^+ balance) in the management of severe hyperkalemia?

❏ Achievement of a net K^+ loss is the most effective therapeutic modality to reduce and sustain a normal $[K^+]_p$. The first two modalities in the therapy of hyperkalemia, namely the use of Ca^{++} salts (to ameliorate the effect of hyperkalemia on the excitability of myocardial and skeletal muscle), and the administration of insulin and glucose, $NaHCO_3$ and albuterol (to modify internal K^+ balance promoting K^+ translocation from ECF to ICF) are only temporary measures that remove the immediate threat to life resulting from hyperkalemia. Consequently, treatment of severe hyperkalemia must combine all three modalities.

416 Describe procedures for K^+ removal (negative external K^+ balance) in the therapy of hyperkalemia.

❑ The administration of polystyrene sulfonate resins (Kayexalate) and the performance of dialysis (hemodialysis and peritoneal dialysis) are effective measures for K^+ removal in patients with hyperkalemia.

417 Describe the use of Kayexalate in the treatment of hyperkalemia.

❑ Kayexalate can be administered orally or in a retention enema (the latter is a more effective route). Optimal action of this resin requires that sorbitol be given concomitantly. A satisfactory formulation for its use in both routes is that of 25 grams of resin, 25 grams of sorbitol, and 100 ml of water. Oral and rectal routes can be used concomitantly, and similar doses can be administered multiple times, even hourly. Kayexalate exchanges Na^+ initially bound to the resin for K^+ initially present in gastrointestinal fluids. Each gram of resin can induce a net K^+ loss of ~ 1 meq.

418 Describe specific reasons for selecting a particular treatment modality, or a combination of them, in the management of severe and/or symptomatic hyperkalemia.

❑ The presence of associated clinical and laboratory abnormalities can prevent use of one or more treatment modalities of hyperkalemia. The use of K^+-exchange resins by oral or rectal routes is contraindicated in patients with significant gastrointestinal symptoms. Calcium infusions are contraindicated in patients with hypercalcemia. Sodium bicarbonate infusions are contraindicated in patients with alkalemia, with high $[HCO_3^-]_p$, with hypernatremia, in those at a significant risk of developing pulmonary edema, and in those with significant ECF volume expansion.

419 Explain the need for dialysis in the treatment of severe hyperkalemia.

❑ Severe hyperkalemia accompanied by a preserved renal function can usually be corrected without dialysis. Potassium removal can be achieved when renal function is preserved by inducing an enhanced kaliuresis with the administration of fluids containing NaCl and/or $NaHCO_3$, and with the use of diuretics (acting on the proximal tubule, loop of Henle, and distal tubule, such as acetazolamide, furosemide, and thiazides, respectively). The majority of patients that develop severe hyperkalemia, however, have renal failure and dialysis is the treatment indicated for this electrolyte abnormality.

420 Explain the use of hemodialysis and peritoneal dialysis in the treatment of severe hyperkalemia.

❑ Both dialysis modalities can successfully control hyperkalemia. However, hemodialysis is substantially more effective than peritoneal dialysis in rapidly achieving a large K^+ removal from body fluids.

421 Calculate the K^+ removal with hemodialysis.

❑ Let us consider a four-hour hemodialysis treatment, with a blood flow through the dialyzer of 200 ml/min and a K^+-free dialysate flowing at 500 ml/min. Assuming a 38% hematocrit (62% plasma), the plasma flow in the four-hour treatment would amount to ~30 liters (240 min x 124 ml/min plasma flow). Since the plasma flow and the dialysate flow in the hemodialysis device have opposite directions, the plasma loses most of its K^+ content upon equilibration with the dialysate. Each liter of plasma that flows through the dialysate can lose, for example, 4 meq of K^+. Thus, the total K^+ removal in the four-hour period would be ~120 meq. Consequently, while the $[K^+]$ in the fresh dialysate was zero meq/liter, the $[K^+]$ in the used dialysate would be ~1 meq/liter (since the 120 liters of used dialysate should contain the 120 meq of K^+ removed from the patient's blood).

422 Calculate the K^+ removal with peritoneal dialysis.

❑ Let us consider a four-hour peritoneal dialysis treatment with two liters per dialysis cycle, and hourly exchanges of K^+-free dialysate. Assuming a proper equilibration of $[K^+]$ in dialysate and plasma, the total K^+ removal would be ~40 meq (5 meq/liter is the $[K^+]$ in used dialysate, multiplied by 8 liters of total dialysate for the period). The calculations made show that, over the same period, the K^+ removal with hemodialysis is approximately three times greater than that obtained with peritoneal dialysis.

423 Describe conditions leading to hyperkalemia, due to a skeletal muscle disease, that respond to a specific therapy.

❏ Patients with hyperkalemic periodic paralysis can be successfully managed with albuterol inhalation and/or oral intake of acetazolamide and/or thiazides. Patients with large areas of skeletal muscle necrosis and/or large hematomas are at risk of hyperkalemia, and debridement of affected areas helps to control the elevated $[K^+]_p$. The hyperkalemia observed in malignant hyperthermia can be prevented/treated with the administration of nandrolone.

424 What is the management of hyperkalemia in Addison's disease?

❏ Patients with Addison's disease (chronic adrenocortical insufficiency) restore their K^+ homeostasis with the administration of fludrocortisone (a synthetic mineralocorticosteroid-acting drug), in addition to glucocorticosteroid therapy.

425 Describe the precipitating factors and management of primary hyperkalemic periodic paralysis.

❏ A variety of stimuli including cold exposure, fasting, ingestion of small amounts of dietary K^+, excitement, rest after intense exercise, and general anesthesia, trigger the crisis of paresis/paralysis. Regarding the management of this disease, an acute attack in its early phase can be prevented by the inhalation of a beta-adrenergic agonist (albuterol) and the administration of carbohydrates (orally and intravenously). Severe acute attacks require the infusion of insulin, HCO_3^-, and Ca^{++}, to counteract the effects of hyperkalemia on skeletal muscle and on the heart. The long term plan aimed at prevention of paralytic attacks includes administration of either thiazides or acetazolamide.

426 Let us review a clinical history. A 63-year-old man was found on the street having a generalized seizure and was brought to the hospital emergency room. Previous history was significant for heavy alcohol (ethanol) consumption and toxic hepatitis (alcoholic) two years prior to this episode. Initial physical examination revealed recurrent generalized seizures; rectal temperature, 38.5° C; blood pressure, 115/55 mmHg; heart rate, 44/min, and respiratory rate, 20/min. The remainder of the examination was within normal limits including absence of meningeal signs and focal neurological deficit. The composition of cerebrospinal fluid was normal and urinalysis showed 4+ hemoglobin, 0 to 2 red blood cells and > 20 white blood cells per high-power field. Laboratory data were as follows:

plasma creatinine	3.0 mg/dl
BUN	18 mg/dl

Serum electrolytes:

$[Na^+]$	128 meq/liter
$[K^+]$	9.1 meq/liter
$[Cl^-]$	98 meq/liter
total CO_2 (T_{CO_2})	8 mmol/liter
anion gap*	22 meq/liter

*$[Na^+]_p - ([Cl^-]_p + [T_{CO_2}]_p)$

Arterial blood gases:

pH	7.15 ($[H^+]$ ~71 nmol/liter)
P_{CO_2}	24 mmHg
P_{O_2}	78 mmHg

What is the most life-threatening laboratory abnormality observed in the initial evaluation of this patient?

❏ Severe hyperkalemia (9.1 meq/liter) is undoubtedly the biochemical abnormality that imposes the greatest risk of sudden death in this patient. The patient's $[K^+]_p$ is high enough to induce serious cardiac arrhythmias and/or circulatory arrest. The decreased heart rate on initial physical examination suggests the possibility of an idioventricular rhythm (contrasting with the normal sinus rhythm), and this abnormal cardiac rhythm was documented with an ECG tracing. The ECG revealed absence of P waves (auricular standstill), prolonged QRS complexes (intraventricular block), and QRS configuration resembling a sinus wave.

427 What is the most likely pathogenesis for the hyperkalemia observed in this patient (described in the previous question) who was admitted with generalized convulsions?

❏ Considering the severe ECG abnormalities observed in the initial tracing, pseudohyperkalemia can be ruled out, and causes for true hyperkalemia must be considered. The patient's history of heavy alcohol intake is consistent with a low dietary K^+ intake that predisposes him to K^+ depletion and hypokalemia (contrasting with the hyperkalemia observed in our patient). According to his clinical history, neither oral nor parenteral K^+ supplementation, neither drugs that impair renal (angiotensin-converting enzyme inhibitors, K^+-sparing diuretics), nor extrarenal (digitalis, beta-adrenergic blocking agents) K^+ disposal were given to this patient. There is no evidence of generalized renal failure severe enough to cause hyperkalemia. The foregoing criteria led us to formulate the presumptive diagnosis of hyperkalemia due to K^+ redistribution from ICF to ECF. A large K^+ leakage from skeletal muscle due to generalized seizures-induced rhabdomyolysis best explains the observed hyperkalemia. The presumptive diagnosis of rhabdomyolysis was subsequently confirmed by the demonstration of a plasma creatine phosphokinase (CPK) level of 12,400 units. The increased level of plasma creatinine (3.0 mg/dl) as compared to BUN (18 mg/dl) is most consistent with release of muscle creatine and subsequent catabolism of this substance to creatinine, contrasting with the simple retention of creatinine and other waste products caused by renal failure.

428 Explain the determinants of the deranged laboratory data observed in this patient with hyperkalemia due to rhabdomyolysis.

❏ The arterial blood gases reveal moderately severe acidemia due to metabolic acidosis of the high anion gap variety. This acid-base disturbance is best explained by the development of lactic acidosis due to excessive anaerobic metabolism in the skeletal muscle subjected to repeated generalized seizures. The hyperkalemia observed in this patient is not explained by the presence of lactic acidosis, since this organic acid-metabolic acidosis does not lead to significant hyperkalemia. The patient's hyponatremia is partly caused by cellular Na^+ entry in exchange for H^+. The presence of heme pigments in the urine as determined by the 4+ hemoglobin, and 0 to 2 red blood cells, is consistent with rhabdomyolysis; the initial finding of 4+ hemoglobin was subsequently found to be largely due to urinary excretion of myoglobin.

429 Describe what should the initial management of this patient's condition be with respect to the hyperkalemia and other clinical/laboratory abnormalities.

❏ Immediate attention should be focused on prevention of further episodes of generalized seizures and to secure adequate pulmonary ventilation. Administration of anticonvulsant medication, intubation of the airway, and initiation of mechanical ventilation should be performed at once. Adequate alveolar ventilation is seriously threatened by seizure activity that disrupts the normal inspiration/expiration cycle and by profound hyperkalemia that can lead to paralysis of respiratory muscles.

Therapy of this patient's hyperkalemia should include agents aimed at ameliorating the cardiac and skeletal muscle toxicity of high $[K^+]_p$, redistributing the excessive levels of extracellular K^+ to the cellular compartment, and facilitating excretion of K^+ by renal and extrarenal routes. The i.v. administration of Ca^{++} salts will have an almost immediate effect to help in correcting abnormal tissue excitability. Insulin and glucose administration will promote the translocation of K^+ from ECF to ICF. Exchange resins such as Kayexalate can be given by enema to remove K^+ by an extrarenal route. Sodium bicarbonate administration will help correction of the hyperkalemia by enhancing renal K^+ excretion as well as by other mechanisms.

Since rhabdomyolysis creates a major risk of acute renal failure, immediate measures should be taken to prevent it from developing, or to correct it if it were already present.

430 Let us review another case. A 22-year-old fireman was severely injured while attempting to control a large petrochemical blaze. The patient was rushed to the emergency room (ER) of a small community hospital, arriving in circulatory collapse with third degree burns covering 60% of his body surface. Once his hemodynamic status had been stabilized with the administration of isotonic crystalloid and colloid i.v. infusion, the patient was transferred to a regional trauma center where he arrived 12 hours later with the following laboratory data:

plasma creatinine	3.3 mg/dl
BUN	20 mg/dl

Serum electrolytes:

$[Na^+]$	129 meq/liter
$[K^+]$	8.5 meq/liter
$[Cl^-]$	99 meq/liter
total CO_2 (TCO_2)	16 mmol/liter

The ECG revealed changes consistent with severe K^+ intoxication. An indwelling bladder catheter had been inserted upon arrival to the ER where the patient received initial treatment but there had been no urine output during the subsequent 12 hours.

Is renal failure-induced hyperkalemia the most likely explanation for the increased $[K^+]_p$ in this patient?

What is the pathogenesis of the increased $[K^+]_p$?

❑ Renal failure is definitely not the major cause of hyperkalemia in this patient. Persistent anuria in spite of restoration of hemodynamic status is indicative of severe acute renal failure. Yet, the absence of effective renal function in a twelve hour period should result in only a very modestly increased $[K^+]_p$ (< 1 meq/liter). Consequently, another explanation must be found to explain the hyperkalemia. The patient's $[K^+]_p$ has risen due to massive tissue death secondary to extensive tissue burning with the release of large amounts of K^+ from ICF to ECF. The hyperkalemia observed in this patient is not caused by increased body K^+ stores but by abnormal K^+ internal redistribution.

431 How should the hyperkalemia in this patient be treated?

❑ The severe hyperkalemia observed in this patient requires a multiple pronged therapy to ensure recovery from this life-threatening electrolyte abnormality. Consequently, administration of calcium gluconate, glucose and insulin, and $NaHCO_3$ by the intravenous route are all indicated. Kayexalate (exchange resin) therapy aimed at removing body K^+ is also indicated if technically possible. In addition, efforts should be made to start dialysis promptly, since other measures are only temporary and have limited value in patients with extensive tissue trauma and with lack of renal function (anuria).

432 Let us review another clinical history. A 78-year-old man was admitted to the general medicine service with complaints of fatigability, poor appetite, weight loss of ~5 kg over the previous six months, and persistent itching. The patient's past medical history was remarkable for labile hypertension, a subendocardial myocardial infarction five years earlier, and nocturia. Upon further questioning, the patient acknowledged difficulties with starting urination as well as completing this process without persistent dribbling of urine. The patient had not been taking any medication, either prescribed or over-the-counter.

Physical examination revealed a healthy-appearing elderly man. Examination of vital signs revealed blood pressure, 170/82 mmHg; pulse, 74/min; and respirations, 26/min. Examination of the skin showed normal turgor and lesions due to trauma induced by scratching in response to persistent itching; these skin lesions were predominantly located in the upper extremities and on the back, and were accompanied by areas of ecchymosis. Body weight was 78 kg. Urinalysis revealed: specific gravity, 1.008; pH, 6.0; glucose, ketones, and proteins, were negative. ECG showed mild left ventricular hypertrophy, but was otherwise unremarkable. Other laboratory data showed:

Hematocrit	40 %
plasma creatinine	4.2 mg/dl
BUN	86 mg/dl

Serum electrolytes:

$[Na^+]$	134 meq/liter
$[K^+]$	6.8 meq/liter
$[Cl^-]$	96 meq/liter

total CO_2 (TCO_2)　　18 mmol/liter
anion gap*　　　　　　　20 meq/liter
　*$[Na^+]_p - ([Cl^-]_p - [TCO_2]_p)$

Provide a general assessment of the patient's main medical problems.

❑ The patient's history, physical examination and laboratory data reveal renal insufficiency as the entity responsible for the exhibited signs and symptoms. The decreased appetite, moderate weight loss, persistent itching, ecchymosis, and nocturia are common manifestations of renal insufficiency.

433 What is the most likely explanation for the hyperkalemia and renal insufficiency observed in this elderly man?

❑ Hyperkalemia of moderate severity (6.8 meq/liter) is properly explained by renal insufficiency. It should be noted, however, that the degree of renal insufficiency, assessed by the plasma creatinine level (4.2 mg/dl), is commonly associated with normal $[K^+]_p$; however, the substantially elevated BUN (86 mg/dl) indicates a major retention of nitrogen waste products, and is therefore compatible with the development of hyperkalemia due to K^+ retention.

The most likely explanation for the patient's renal insufficiency is acute/subacute obstructive uropathy based on the following facts: (1) the patient is an elderly male with a negative history of previous renal disease and with normal hematocrit; these elements indicate that renal insufficiency was most likely of an acute/subacute nature; (2) urinalysis results that are negative for proteins, cells, and casts, suggesting that an intrinsic renal disease is an unlikely cause for the observed renal insufficiency, favoring instead, an extrarenal mechanism (prerenal or postrenal); since the patient had had no evidence of ECF volume depletion or hypotension that could lead to prerenal azotemia, the most likely explanation for the patient's renal insufficiency is a postrenal mechanism; (3) the abnormal micturition reported by this elderly male suggests that the postrenal mechanism responsible for hyperkalemia and azotemia is obstructive uropathy at the level of the urethra and/or the urinary bladder. Subsequent studies confirmed the presumptive diagnosis of obstructive uropathy, which had been caused by a benign enlargement of the prostate.

434 Describe the management of hyperkalemia and renal insufficiency observed in this elderly man.

❑ When adequate urine flow is re-established in a patient with obstructive uropathy, hyperkalemia and azotemia promptly recede. Since toxic cardiac effects of hyperkalemia were absent, emergency treatment of hyperkalemia was not indicated (i.e., i.v. calcium infusions, glucose and insulin administration, dialysis). Insertion of an indwelling bladder catheter and adequate fluid intake rapidly improved the patient's condition.

Twelve hours later, the laboratory data showed:

Hematocrit	38 %
plasma creatinine	3.6 mg/dl
BUN	60 mg/dl

Serum electrolytes:

$[Na^+]$	138 meq/liter
$[K^+]$	4.6 meq/liter
$[Cl^-]$	98 meq/liter
total CO_2 (TCO_2)	24 mmol/liter
anion gap*	16 meq/liter

*$[Na^+]_p - ([Cl^-]_p + [TCO_2]_p)$

Medical intervention with bladder catheterization and provision of fluids served two purposes: (1) it corrected hyperkalemia and azotemia; and (2) it established with certainty that obstructive uropathy was responsible for the clinical syndrome and laboratory abnormalities observed on admission.

435 Let us review another case. A 14-year-old boy was admitted to the hospital because of severe abdominal pain and vomiting of six hours duration. The diagnosis of acute appendicitis was made and he was brought to the operating room for surgical treatment. Routine laboratory data including serum electrolytes, obtained before surgery, were within normal limits. Minutes after the initiation of general anesthesia with halothane and succinylcholine, the patient developed muscle fasciculations, rigidity, tachycardia, and hypotension. Body temperature rose rapidly from a preoperative value of 37.2° C to 41.5° C; the observed high fever was initially associated with warm and red skin that changed to cold skin with patchy cyanosis most evident in the extremities. Laboratory data revealed:

$[K^+]_p$	7.7 meq/liter
$[Mg^{++}]_p$	4.2 meq/liter
serum lactate	108 mg/dl (12 meq/liter)

Arterial blood gases:

pH	6.92 ($[H^+]$ ~120 nmol/liter)
P_{CO_2}	18 mmHg
$[HCO_3^-]_p$	4.0 meq/liter

What is your interpretation of the acute illness and hyperkalemia experienced by the patient during surgery?

❑ The acute syndrome developed by this child, although unusual, is typical of the entity known as malignant hyperthermia. Potassium released from skeletal muscle in patients with malignant hyperthermia is responsible for the development of hyperkalemia.

436 What are the most salient characteristics of malignant hyperthermia?

❑ Malignant hyperthermia encompasses several inherited disorders characterized by the acute onset of hyperthermia due to muscle contractions (patients develop fasciculations, trismus, and overall muscle rigidity), accompanied by tachycardia and hypotension, when exposed to several inhalant anesthetics (halothane, methoxyflurane, cyclopropane, ethyl ether) and/or muscle relaxants (succinylcholine). Laboratory data revealed the presence of hyperkalemia, hypermagnesemia, and lactic acidosis in this child. This syndrome can result in massive skeletal muscle swelling due to rhabdomyolysis of severe degree, and life-threatening complications (disseminated intravascular coagulation, pulmonary edema, acute renal failure).

437 Describe the pathogenesis of malignant hyperthermia and the possible relationship with so-called kaliogenic hypermetabolism.

❑ Inhalant anesthetics and depolarizing drugs trigger the release of Ca^{++} in sarcoplasmic reticulum of muscle cells and induce an abnormal cycle of excitation/contraction leading to the release of heat, K^+, and other cellular constituents. Increased heat production leading to hyperthermia and hyperkalemia (similar to that found in malignant hyperthermia) develops when skeletal muscle of experimental animals is exposed to high $[K^+]_e$. This process, known as kaliogenic hypermetabolism, is characterized by stimulation of glucose consumption, and production of heat, CO_2, and lactate due to Ca^{++} release within muscle cells. The effect of $[K^+]_e$ on skeletal muscle metabolism and heat production is direct and linear, producing a large increment of these parameters in response to hyperkalemia.

438 What is the management of hyperthermia and hyperkalemia caused by severe rhabdomyolysis in patients with malignant hyperthermia?

❏ Since untreated attacks of malignant hyperthermia are usually fatal, this entity represents a medical emergency. The following measures must be immediately implemented upon recognition of the syndrome: (1) discontinue administration of the anesthetic/muscle relaxant that triggered the crisis; (2) administer 100% O_2; (3) provide external cooling; (4) administer a rapid i.v. infusion of 1 to 10 mg/kg dantrolene sodium, a specific drug for the control of this syndrome; and (5) provide i.v. fluids containing $NaHCO_3$ and diuretics (e.g., furosemide) to control hyperkalemia and metabolic acidosis and to diminish the risk of acute renal failure due to rhabdomyolysis.

Selected References

1. Adrogué HJ (ed): Acid-base and electrolyte disorders. Contemporary Management in Critical Care, 1:2. Churchill Livingstone, New York, 1991.

2. Hatano M (ed): Nephrology. Springer-Verlag, Tokyo, 1991.

3. Seldin DW and Giebisch G (eds): The Regulation of Potassium Balance. Raven Press, New York, 1989.

4. Tannen RL (ed): Fluid, Electrolyte and Acid-Base Disorders. In: The Principles and Practice of Nephrology. Jacobson HR, Striker GE, and Klahr S (eds). B.C. Decker, Inc., Philadelphia, 1991.

5. Whelton PK, Whelton A, and Walker WG (eds): Potassium in Cardiovascular and Renal Medicine. Marcel Dekker, Inc., New York, 1986.

Index

Numbers refer to questions

Acetazolamide 161, 162, 240, 265, 305, 398
Acid-base balance *(see acid-base disturbances)*
Acid-base disorders *(see acid-base disturbances)*
  potassium depletion and metabolic alkalosis 178-181, 202
Acid-base disturbances and plasma potassium concentration ($[K^+]_p$)
  acidosis 124-130, 133, 135-140, 372, 403
  alkalosis 124-136
  diabetic ketoacidosis 300-303
  lactic acidosis 426-429, 435-438
  metabolic acidosis 131-141, 262, 263, 265, 371, 372, 403
  metabolic alkalosis 131-136, 211, 262, 263
  renal tubular acidosis 265, 309-312, 371, 372, 403
  respiratory acidosis 135, 136
  respiratory alkalosis 132, 135, 136, 211
Acid-base disturbances and potassium excretion
  acidosis 178-180, 372, 403
  alkalosis 178-181
  diabetic ketoacidosis 300-303
  metabolic acidosis 178-181, 265, 309-312, 371, 372, 403
  metabolic alkalosis 178-181
  respiratory acidosis 178-181
  respiratory alkalosis 178-181
  uremic acidosis 178-180, 182, 184-186, 364, 432-434
Acid excretion, renal
  in potassium deficiency 202
Acidosis *(see acid-base balance and disturbances)*
ACTH (adrenocorticotropic hormone) 325, 326, 374, 376
Addison's disease 169, 170, 424
Adenoma, adrenal *(see adrenal adenoma)*
  colon, villous and hypokalemia 234, 238

ADH *(see antidiuretic hormone)*
Adrenal adenoma and hyperplasia
  hypokalemia 158, 159, 167-171, 297, 325, 326
  primary aldosteronism 158, 159, 167-171, 297, 325, 326
Adrenal enzyme defects 326
Adrenocortical insufficiency *(see Addison's disease)*
Adrenocorticotropic hormone (ACTH) 325, 326, 374, 376
Alcohol and potassium homeostasis 228, 321-324, 426-429
Aldosterone *(see also renin-angiotensin-aldosterone axis)*
  colon, effects on 234, 235, 397
  deficiency of 296, 297, 364, 367-379, 403, 404
  diuretic-induced high levels 244
  hypersecretion *(see aldosteronism)*
  hypokalemia-induced 158, 159, 167-171, 297, 325, 326
  potassium secretion 158, 159, 164-171, 297, 326, 372, 403
  production, excessive *(see aldosteronism)*
  pseudohypoaldosteronism 372, 384, 386
  secretion, in Bartter's syndrome 245
  spironolactone effects 293-297
Aldosteronism *(see also aldosterone, adrenal adenoma, and hyperplasia)*
  amiloride 160, 293-299
  diuretic-induced 244
  metabolic alkalosis in 262, 263, 267, 325, 326
  primary 167-171, 325, 326
  pseudoaldosteronism 384, 386
  secondary 167-171, 244, 325, 326
  spironolactone 293-297
  triamterene 293-299
Alkalemia and plasma potassium concentration ($[K^+]_p$) *(see acid-base disturbances and plasma potassium concentration)*
Alkali therapy in hyperkalemia 124-136, 403, 410, 411

Alkalosis *(see acid-base balance and disturbances)*
Amiloride *(see diuretics)*
Ammonia
   excretion in urine 202
   hepatic encephalopathy 228, 279, 280, 321-325
   in potassium depletion 202
   production of, and potassium homeostasis 202
Ammoniagenesis *(see ammonia)*
Anesthesia
   hyperkalemia 347, 351, 352, 435-438
   malignant hyperthermia 347, 351, 352, 435-438
Angiotensin *(see renin-angiotensin-aldosterone axis)*
Angiotensin converting enzyme inhibitor *(see converting enzyme inhibitors)*
Anions
   cell-restricted *(see also Gibbs-Donnan Rule)* 16-20, 24, 52
   diffusible 16, 19, 24-27, 36
   extracellular 16, 19, 24, 26, 27
   intracellular 16, 18-20
Anorexia nervosa and hypokalemia 230
Antidiuretic hormone (ADH)
   kaliuresis, effects on 172, 175-177, 191, 202
Arrhythmias *(see cardiac arrhythmias)*
Asterixis *(see flapping tremor)*
Asthma *(see stress hypokalemia)*
ATPase
   H^+, K^+-ATPase 259
   Na^+, K^+-ATPase 28-30, 46, 54-61, 65, 76, 78, 113-115
Azotemia *(see renal failure)*
Barium poisoning, redistribution of potassium and 90, 91, 212, 219, 306
   hypokalemia 90, 91, 212, 306
Bartter's syndrome 245, 246, 304
Beta-adrenergic agonists and antagonists *(see catecholamines)*
Bicarbonate
   concentration in plasma and $[K^+]_p$ 124-130, 403, 410, 411
   urine concentration and renal potassium secretion 152-155, 178-181, 239-241

Bicarbonate therapy *(see alkali therapy)*
Blood pressure
   dietary potassium, effects on 191, 193-196
   hypokalemia 191, 193-196, 198, 202
   racial differences 196
   socioeconomic status 196
Bone, potassium content 13
Brain edema 61, 65
Calcium
   administration of, for hyperkalemia 100-104, 405-407, 412
   cardiac excitability 100-104, 313, 314
   concentration in body fluids 10
   neuromuscular excitability 100-104, 313, 314
   threshold potential, effects on 100-104, 313, 314, 332
Captopril *(see converting enzyme inhibitors)*
Carbohydrate metabolism, potassium depletion and 1, 191, 203
Carbonic anhydrase inhibition *(see acetazolamide)*
Cardiac arrhythmias
   and plasma potassium, concentration 192, 193, 278, 287, 313, 314, 330-336
   digitalis 28, 56, 193, 345, 347, 349, 350
   diuretics *(see diuretics)*
Catecholamines
   alpha adrenergic agonists and antagonists 111, 118, 119, 121-123, 204, 206, 347
   asthma and hypokalemia *(see stress hypokalemia)*
   beta-adrenergic agonists and antagonists 111, 118, 120-123, 206-209, 243, 253-255, 344, 345, 347, 413, 414, 423
   chronic obstructive pulmonary disease and hypokalemia *(see stress hypokalemia)*
   epinephrine *(see catecholamines)*
Cation exchange resin, in hyperkalemia *(see Kayexalate)*
Cations
   diffusible *(see also Gibbs-Donnan rule)* 16, 19, 20, 23, 25-27, 36, 38
   extracellular 9, 10, 22
   intracellular 7, 8

Channels, potassium
  barium effects 90, 91, 212, 219
  inhibitors 88, 90-94
  insulin effects 113, 114, 116
  insulin secretion 92-95
  oral hypoglycemic agents 92-95
  properties 75, 78-86, 89-94
  types 86-94
Chemotherapy, and hyperkalemia 354, 359
Chloride
  distribution in body fluids 19, 24, 26, 27, 31
  shunt (see also Gordon's syndrome) 386
  urine concentration in potassium excretion 152, 155, 174, 177, 239, 240
Chronic obstructive lung disease and plasma potassium concentration (see stress hypokalemia)
Cirrhosis, hepatic encephalopathy and potassium depletion 191, 203, 228, 279, 280, 321-325
Clay ingestion (geophagia) and potassium loss 226, 229
Collecting tubule, potassium transport in 150-186, 265, 309-312, 372, 403
Colon, sodium absorption and potassium secretion in 147, 165-168, 171, 231, 232, 234, 238, 397
Coma, hepatic and potassium homeostasis (see hepatic coma)
Conductance 62, 64, 77, 80-83
Converting enzyme inhibitors (ACE inhibitors) (see also renin-angiotensin-aldosterone axis)
  Bartter's syndrome 304
  treatment of hypokalemia and risk of hyperkalemia 277, 281, 292-296, 304, 307, 348, 373, 376, 379, 402
Corticosteroids
  glucocorticosteroids 169, 170, 222, 223, 267, 326, 376, 379, 424
  mineralocorticosteroids 153-160, 164-171, 241, 244, 245, 267, 296, 297, 325, 326, 364, 367-379, 386, 403
  renin-angiotensin-aldosterone axis 368-379
Corticosterone (see corticosteroids)
Cortisol (see corticosteroids)

Cushing's syndrome (see corticosteroids, glucocorticosteroids)
Cyclosporine and plasma potassium levels 348, 379-381
Deoxycorticosterone (see corticosteroids)
Diabetes insipidus, nephrogenic (see also ADH)
  in potassium depletion 161, 172, 175-177, 191, 202
Diabetes mellitus (see also insulin)
  hyperkalemia 296, 371, 372, 403
  hypokalemia 203, 210, 300-303, 320
  hyporeninemic hypoaldosteronism 296, 371, 372, 403
  potassium channels in 92-95
Diabetic ketoacidosis
  and plasma potassium concentration ($[K^+]_p$) 210, 300-303, 320
  potassium replacement therapy 210, 300-303, 320
Dialysis
  hemodialysis, K^+ removal 416, 419, 421
  peritoneal dialysis, K^+ removal 416, 419, 420, 422
  therapy of hyperkalemia 416, 419-422
Diarrhea and potassium excretion 146, 147, 231-238, 315-319, 397
Diet
  potassium content in various foods 393-396
  potassium intake 109, 144, 145
Digitalis
  action 28, 61
  plasma potassium levels 345, 347, 349, 350, 358
  sodium, potassium-ATPase (Na^+, K^+-ATPase) 28, 55, 56, 61, 349, 350
  toxicity 345, 347, 349, 350, 358
Distal nephron 150-186, 309-312, 371, 372, 403
Diuretics
  acetazolamide 161, 162, 240, 265, 305, 398
  aldosterone (see aldosterone, aldosteronism)
  cardiac arrhythmias and (see cardiac arrhythmias)
  digitalis (see digitalis)

Diuretics *(continued)*
  furosemide 161, 162, 240, 242-244, 260, 269, 278, 280, 398, 403
  hyperkalemia *(see diuretics, potassium-sparing)*
  hypokalemia 242, 243, 260, 280, 323
  loop diuretics *(see diuretics, furosemide)*
  mannitol 142, 243, 353, 412
  osmotic 142, 243, 353, 412
  potassium-sparing
    aldactazide 293-299
    amiloride 160, 293-299
    dyazide 161, 162, 293-299
    maxzide 162, 293-299
    spironolactone 293-297
    triamterene 293-299
  renin-angiotensin-aldosterone axis *(see renin-angiotensin-aldosterone axis)*
  thiazide diuretics 161, 162, 240, 242-244, 260, 269, 278, 280, 293-299, 398, 403
Duct, collecting 150-186, 265, 309-312, 372, 403
ECF *(see extracellular fluid)*
ECG (electrocardiogram)
  in hyperkalemia 330, 331, 333-336, 342
  in hypokalemia 192, 193, 249, 331
Edema, brain 61, 65
EKG *(see ECG)*
Enalapril *(see converting enzyme –ACE– inhibitors)*
End-organ resistance to aldosterone 372, 384, 403
Epinephrine *(see catecholamines)*
Erythrocyte *(see red cell)*
Ethacrynic acid *(see diuretics, loop diuretics)*
Evolution of life forms 5, 6
Extracellular fluid (ECF)
  comparison with ICF 7, 9-11, 14-16
  electrolyte composition 9, 10, 22, 24
  normal potassium level 105-107
  tonicity and plasma potassium concentration 111, 142
Flapping tremor
  hepatic encephalopathy 203, 279, 280, 321-325
Fractional excretion of potassium in the urine 182-186
Furosemide *(see diuretics)*

Gastric fluid
  electrolyte composition 232, 233
  loss of, in metabolic alkalosis 233, 325, 326
Gastrointestinal
  motility and potassium disorders 197, 337
  loss of potassium 146, 147, 231-238, 315-319, 397
Geophagia (clay ingestion) 226, 229
Gibbs-Donnan rule 15-17, 25-29, 52
Glucagon 117, 344, 356
Glucocorticoid hormones *(see corticosteroids, glucocorticosteroids)*
  action compared with mineralocorticosteroid hormones 169, 170
Glucose
  hyperglycemia-induced hyperkalemia 142, 353
  intolerance with potassium depletion 1, 92-95, 203
Glycyrrhizic acid (licorice ingestion) 267, 325, 326
Goldman, Hodgkin, Katz equation 71
Gordon's syndrome (chloride shunt) 386
Heart *(see cardiac arrhythmias)*
Hemodialysis *(see dialysis)*
Hemolysis and hyperkalemia *(see also red cell)* 72, 344, 347, 353, 359
Heparin and plasma potassium concentration ($[K^+]_p$) 379, 383
Hepatic *(see also liver)*
  coma, in hypokalemia 191, 203, 228, 279, 280, 321-325
  encephalopathy, in potassium depletion 228, 279, 280, 321-325
High blood pressure *(see blood pressure)*
Hormones, effect on potassium levels *(see ADH, catecholamines, corticosteroids, glucagon, and insulin)*
Hydrogen ion(s)
  and potassium distribution 59, 113, 115, 124-140
  concentration of, and plasma potassium levels 124-140
  H^+, K^+-ATPase 259
  Na^+, H^+ exchanger 59, 113, 115
Hydronephrosis and hyperkalemia 372, 434

Hyperaldosteronism *(see aldosteronism)*
Hypercalcemia
  and cardiac excitability 101, 102,
    104, 313, 314
  and neuromuscular excitability 101,
    102, 104, 313, 314
Hypercapnia and potassium levels *(see
  acid-base disturbances and $[K^+]_p$)*
Hyperglycemia and plasma potassium
  concentration 142, 353, 408
Hyperkalemia *(see also potassium)*
  acid-base disturbances *(see acid-base
    disturbances)*
  action potential 330, 331, 333
  acute renal failure *(see renal failure)*
  adrenal insufficiency and 169, 170,
    296, 297, 364, 368-379, 424
  anesthesia 347, 351, 352, 435-438
  arrhythmias in *(see cardiac arrhythmias)*
  artifactual, pseudo, false, or spurious
    328
  associated with
    insulin deficiency *(see diabetes,
      diabetic ketoacidosis)*
    potassium retention 364-389
    potassium redistribution
      343-359
  bicarbonate therapy 124-136, 408,
    409
  blood transfusions 361
  calcium therapy 100-104, 405-407,
    412
  cation exchange resins *(see also
    Kayexalate)* 416, 417
  causes 343-386, 400, 403, 423
  clinical approach 342-387
  clinical manifestations 329-341
  degrees of severity 342
  definition 327
  diabetes mellitus/diabetic ketoacidosis
    296, 300-303, 344, 345, 371
  dietary treatment 391-396
  diuretics *(see diuretics)*
  drug-induced 344-354
  ECG abnormalities 330, 335, 336,
    342
  effects 99, 101, 329-341
  electrocardiography *(see ECG)*
  endocrine/metabolic effects 145, 329,
    341

Hyperkalemia *(continued)*
  gastrointestinal and urinary motility,
    effects on 197, 337
  glucose, insulin, and sodium bicar-
    bonate "cocktail" for 408, 412, 418
  hemodialysis therapy 416, 419-421
  hyperosmolality 142, 353, 412
  hypertonic sodium 142, 353, 412
  hypoaldosteronism *(see also aldoste-
    rone)* 364, 367-379, 403
  hypothermia 208
  in hemolysis 72, 344, 347, 353, 359
  in metabolic acidosis 124-129,
    135-140, 372, 403
  in renal failure 184-186, 364-389,
    399, 419-422, 430-434
  in renal tubular acidosis (RTA) 371,
    372, 403
  in respiratory acidosis 129, 135, 136
  in rhabdomyolysis 347, 359, 426-429,
    435-438
  in uremic acidosis 124-129, 135-140,
    184-186, 364-389, 399, 403,
    419-422, 430-434
  insulin therapy 408, 409, 414
  kayexalate 416, 417
  malignant hyperthermia 423,
    435-438
  management 391-425
  mannitol 142, 353, 412
  membrane potential 99
  mineralcorticoid-resistant 371, 372,
    384, 386, 403
  muscle weakness *(see hyperkalemia,
    neuromuscular effects)*
  neuromuscular effects 329, 331, 332,
    338, 339, 355, 356, 423, 425
  paresis/paralysis of skeletal muscle
    (see hyperkalemia, neuromuscular effects)
  pathogenesis 343-386
  periodic paralysis 338, 355, 356, 423,
    425
  potassium-sparing diuretics *(see diuretics)*
  potassium supplementation 268-277,
    362, 363, 392-396
  pseudo, false, artifactual, or spurious
    hyperkalemia 328
  quadriplegia *(see hyperkalemia, neuromus-
    cular effects)*
  redistribution 343-359

Hyperkalemia *(continued)*
  renal effects of 340
  renal transplantation 348, 372, 379-381, 384
  risk of, in K^+ replacement 276, 277
  sodium chloride restriction, and 400
  sympathetic nervous system 118, 119, 121, 123
  therapy for 391-425
  treatment *(see therapy)*
Hypernatremia *(see also sodium)* 142, 353, 412
Hyperosmolality 142, 353, 408, 412
Hyperplasia, adrenal *(see adrenal adenoma and hyperplasia)*
Hyperreninemia *(see renin-angiotensin-aldosterone axis)*
Hypertension *(see also high blood pressure)*
  hypokalemia 194-196, 326
  in aldosteronism *(see aldosteronism)*
  in Cushing's syndrome *(see corticosteroids, glucocorticosteroids)*
  renal vascular 326
Hyperventilation, alveolar *(see acid-base disturbances, respiratory alkalosis)*
Hypoaldosteronism *(see also aldosterone, aldosteronism, and renin-angiotensin-aldosterone axis)*
Hypocalcemia *(see calcium)*
  and cardiac excitability 101, 103, 104, 331, 332
  and neuromuscular excitability 101, 103, 104, 331, 332
Hypocapnia *(see acid-base disturbances, respiratory alkalosis)*
Hypokalemia *(see also potassium depletion)*
  acid-base disturbances *(see acid-base disturbances)*
  action potential 70, 97
  acute renal failure *(see renal failure)*
  aldosteronism *(see aldosterone, aldosteronism)*
  associated with
    potassium deficiency 205, 221-248
    potassium redistribution 205-220
  barium intoxication *(see barium)*
  Bartter's syndrome 245, 246, 304
  beta blockers *(see catecholamines)*
  blood pressure *(see blood pressure)*

Hypokalemia *(continued)*
  cardiac arrhythmias *(see cardiac arrhythmias)*
  catecholamines *(see catecholamines)*
  causes 205-214, 221-231, 260-267
  clinical approach 255-267
  definition 190
  degrees of severity 249
  diabetes insipidus, nephrogenic 172, 175-177, 191, 202
  dialysis-induced 421-422
  dietary treatment 268-277, 281-289, 293, 294, 391-396
  differential diagnosis 250, 254-257, 262, 267
  diuretic-induced 161, 162, 240, 242-244, 261, 265-267, 280, 293-299, 403
  drug-induced 161, 162, 240, 265-267, 293-299, 305
  ECG abnormalities 192, 193, 249
  effects 98, 101, 191-204, 222, 223
  electrocardiography *(see ECG)*
  enalapril *(see converting enzyme–ACE–inhibitors)*
  encephalopathy 61, 65, 321-325
  endocrine/metabolic effects 203, 204, 222, 223
  epinephrine-induced *(see catecholamines, stress hypokalemia)*
  gastrointestinal and urinary motility, effects on 197, 337
  glucose metabolism 1, 92-95, 203
  hepatic encephalopathy 61, 65, 321-325
  hypercalcemia 101, 102, 104, 313, 314
  hypocalcemia 101, 103, 104
  hypomagnesemia 247, 248
  hypothermia 208
  ileus 197, 337
  in Cushing's syndrome 222, 223, 267, 326
  in diabetic ketoacidosis 210, 300-303, 320
  in metabolic alkalosis 124-136, 178-181, 211
  in mineralocorticoid excess *(see aldosteronism)*
  in renal tubular acidosis (RTA) 265, 309-312

Hypokalemia *(continued)*
  in respiratory alkalosis 132, 135, 136, 178-181, 211
  incidence 242, 252, 253
  insulin *(see diabetes mellitus, insulin)*
  laxative abuse 231, 234, 235, 238, 246, 315-319
  licorice (glycyrrhizic acid) and 267, 325, 326
  Liddle's syndrome and 267, 326
  magnesium 245, 247, 248, 266, 267
  management 268-312
  membrane potential 98
  methylxanthines *(see stress hypokalemia)*
  muscle *(see hypokalemia, neuromuscular effects)*
  myopathy *(see hypokalemia, neuromuscular effects)*
  nephropathy 191, 202
  neuromuscular effects 191, 197-201, 278, 305, 307-314
  oral replacement therapy 145, 268-277, 281-285, 287-295, 362
  osmotic diuresis and 142, 243, 353, 412
  paralysis of skeletal muscle *(see hypokalemia, neuromuscular effects)*
  pathogenesis 205-214, 221-231, 260-267
  periodic paralysis 214-220, 305, 308
  polyuria 161, 172, 175-177, 202, 273, 275, 364-366
  prevalence 252
  prevention *(see potassium repletion)*
  pseudohypokalemia (spurious) 250, 251
  redistribution 205-214
  renal effects 191, 202
  risk of severe manifestations of 278-280
  skin losses 146, 261
  sodium, potassium-ATPase *(see Na^+, K^+-ATPase)*
  spurious (pseudohypokalemia) 250, 251
  stress hypokalemia 253, 255, 257
  sympathetic nervous system *(see also catecholamines)* 118, 120-123
  systemic vascular resistance 191, 193-196

Hypokalemia *(continued)*
  thyrotoxicosis and 216, 308
  total body potassium 12, 13
  treatment *(see also potassium repletion)* 268-312
  urinary retention 197, 337
  urine potassium as diagnostic aid 256-259
  vasoconstriction 193-196
  vomiting and 232, 233, 325, 326
Hypokalemic paralysis 199-201, 214-220, 305, 307-312
Hypomagnesemia 247, 248, 266, 267
Hyporeninemic hypoaldosteronism *(see renin-angiotensin-aldosterone axis)*
Hyporeninism *(see renin-angiotensin-aldosterone axis)*
Hypothermia and $[K^+]_p$ 208
Hypoventilation, alveolar *(see acid-base disturbances, respiratory acidosis)*
ICF *(see intracellular fluid)*
Indomethacin, in Bartter's syndrome 304, 348, 371, 379, 382, 391
Insulin
  and serum potassium concentration 111, 112, 114, 145, 203, 206, 210, 300-303, 320, 372
  cell Ph, effects of 113, 115
  cellular potassium uptake 111-116, 145, 203, 206, 210
  deficiency, and hyperkalemia 344, 353, 372, 408, 409, 414
  diabetic ketoacidosis (DKA) 210, 300-303, 320
  effect of potassium depletion on 92-95
  secretion, defect of 92-95
  sodium, hydrogen counterexchange 59, 113, 115
  sodium, potassium-ATPase (Na^+, K^+-ATPase) 113-115
  therapy, in hyperkalemia 408, 409, 414
Integument *(see skin)*
Intestinal excretion of potassium 146, 147, 231-238, 315-319, 397
Intracellular fluid (ICF)
  electrolyte composition 3, 4, 7, 8, 11, 18, 19, 22-24, 73
  potassium binding capacity 126-135

Intracellular fluid (ICF) *(continued)*
  potassium levels 7, 11, 15, 27, 72
  potential difference *(see potential and potential difference)*
Ischemia of skeletal muscle and hypokalemia 194, 199
Isotopes, estimation of potassium stores 12, 13
Juxtaglomerular apparatus *(see renin-angiotensin-aldosterone axis)*
Kaliuresis
  antidiuretic hormone (ADH), effects on 172, 175-177, 191, 202
  diuretics and *(see diuretics)*
  in mineralocorticoid excess *(see aldosterone and aldosteronism)*
  in potassium-loading 148, 163-165, 167-171
  potassium-sparing diuretics *(see diuretics)*
  regulation 148-189
  sodium chloride (NaCl) restriction, and 400, 403
  urine volume, effects of 161, 172, 175-177, 273-275, 364-366
Kayexalate, therapy for hyperkalemia 403, 416, 417
Ketoacidosis *(see diabetic ketoacidosis)*
Lactic acidosis
  and internal potassium balance 137-140
  and plasma potassium concentration ($[K^+]_p$) 137-140, 426-429, 435-438
Laxative abuse
  comparison with Bartter's syndrome 246
  hypokalemia and *(see also hypokalemia)* 231, 234, 235, 238, 246, 315-319
Licorice (glycyrrhizic acid) ingestion and hypokalemia 267, 325, 326
Liddle's syndrome 267, 326
Liver
  and potassium homeostasis 13, 145
  hepatic encephalopathy 191, 203, 228, 279, 280, 321-325
Magnesium deficiency
  aldosterone levels 245, 247, 248
  relationship to potassium depletion 245, 247, 248, 266, 267
Magnesium, levels in body fluids 8, 10
Malignant hyperthermia 423, 435-438

Mannitol 142, 243, 353, 412
Metabolic acidosis *(see acid-base disturbances)*
Metabolic alkalosis *(see acid-base disturbances)*
Mineralocorticosteroids *(see aldosterone and aldosteronism)*
Muscle disturbances *(see hyperkalemia, neuromuscular effects) (see hypokalemia, neuromuscular effects)*
Muscle relaxants
  hyperkalemia 347, 351, 352, 435-438
  malignant hyperthermia 423, 435-438
Muscle, skeletal *(see skeletal muscle)*
Na^+, K^+-ATPase
  and digitalis 28, 55, 56
  and insulin 113-115
  role of, in cellular levels of sodium and potassium 28-30, 46, 54-61, 65, 76, 78
NaCl in diet and kaliuresis *(see also kaliuresis)* 174, 177, 400
Nephrogenic diabetes insipidus and hypokalemia 202
Nernst equation 39-43, 47, 48, 51
Neuromuscular disturbances in
  hypercalcemia 102, 313, 314
  hyperkalemia 321, 331, 332, 338, 339, 355, 356
  hypocalcemia 101, 103, 104, 331, 332
  hypokalemia 191, 197-201, 313, 314
  hypomagnesemia 245, 247, 248, 266, 267
NH_3 *(see ammonia)*
NH_4^+ (ammonium) *(see ammonia)*
Nitrogen and potassium metabolism 1, 170, 222, 223
Oliguria and potassium excretion *(see kaliuresis)* 161, 172, 175-177, 364-366, 403
Paralysis
  hyperkalemic 329, 331, 332, 338, 339, 355, 356, 423, 425
  hypokalemic 191, 197-201, 214-220, 278, 305, 307-314,
  periodic 214-220, 305, 308
P_{CO_2} *(see acid-base disturbances)*
Periodic paralysis
  causes 212, 216-219
  glucose administration 217, 308

Periodic paralysis *(continued)*
  Goldman-Hodgkin-Katz equation 71
  hyperkalemic 329, 331, 332, 338, 339, 355, 356, 425
  hypokalemic 191, 197-201, 214-220, 278, 305, 307-314
  insulin 217, 308
  management 305, 307, 308, 423, 425
Peritoneal dialysis 416, 419, 420, 422
Permeability 63, 64, 69-71, 77
Ph *(see also acid-base disturbances)*
Phosphate
  intracellular levels 16-18, 21
  intracellular potassium binding capacity 16, 17
P_{K^+} (potassium permeability) 63, 64, 71, 77
P_{Na^+} (sodium permeability) 63, 64, 69-71, 77
Polyuria and potassium excretion *(see also kaliuresis)* 161, 172, 175-177, 364-366, 403
Potassium *(see also hyperkalemia, hypokalemia)*
  acid-base balance and *(see acid-base disturbances)*
  adaptation to K^+ depletion 258, 259
  adaptation to K^+ loading
    extrarenal 165, 166, 397
    renal 165, 182-186, 388, 389
  aldosterone and *(see aldosterone and aldosteronism)*
  and ammoniagenesis 202, 228, 279, 280, 321-325, 372
  and asthma 253, 255, 257
  and cardiac arrhythmias *(see cardiac arrhythmias)*
  and digitalis *(see digitalis)*
  and muscle weakness *(see hyperkalemia and hypokalemia, neuromuscular effects)*
  and renal failure *(see renal failure)*
  balance
    external 109, 110, 143, 144, 360, 361
    internal 110-123, 145, 205-215, 253-255, 357-359
  barium (Ba^{++}) effects 90, 91, 212, 219, 306
  binding capacity of ICF 126-135
  body stores 11-13, 108-110

Potassium *(continued)*
  cellular
    concentration 3, 4, 7, 20, 21, 43, 57, 59, 60, 72, 108
    entry *(see also Na^+, K^+-ATPase)* 76
    exit *(see also channels, potassium)* 75
  cell volume regulation 1, 61, 65
  channels in cell membrane *(see also channels)* 77-94
  concentration of
    extracellular 9, 10, 22, 24, 224
    intracellular 3, 4, 7, 11, 15, 27, 72, 224
  conductance of cell membrane 62, 64, 77, 80-83
  cotransport with chloride 75, 78, 162
  deficiency of *(see hypokalemia)*
  depletion *(see hypokalemia)*
  dietary
    and adaptation of the colon 148, 153, 163, 164, 166, 196
    and adaptation of the distal nephron 148, 153, 163-165, 183, 196
    composition of foods 393-396
    intake 109, 144, 145
  disposal
    extrarenal 110-123, 145, 205-215, 253-255, 357-359
    renal 109, 110, 143, 144, 265, 309-312, 360, 361, 371, 372, 403
  distribution in body fluids 3, 4, 7, 9, 11, 13-16, 26-31, 34, 43, 57, 108
  diuretics and *(see diuretics)*
  effect of mineralocorticosteroids on *(see aldosterone and aldosteronism)*
  equilibrium potential *(see potential and potential difference)*
  formulations *(see potassium repletion)*
  gastrointestinal effects 197, 337
  in Bartter's syndrome 245, 246, 304
  in diabetic ketoacidosis *(see diabetes mellitus and diabetic ketoacidosis)*
  in dialysate 416, 419-422
  in enteral/parenteral feeding 145, 269-277, 281-295, 300-303, 362, 363

Potassium *(continued)*
  in hypertension 191, 193-196
  mineralocorticosteroids *(see aldosterone and aldosteronism)*
  normal concentration
    extracellular fluid (ECF) 105-107
    intracellular fluid (ICF) 3, 4, 7
  ratio to sodium
    fecal 146, 147, 168, 171, 235
    urine 168, 171, 173, 174
  role in biological systems 1, 2
  serum level, comparison with plasma level 106
  supplements 145, 269-277, 281-295, 300-303, 362, 363
  sympathetic nervous system 118-123
  therapy 268-277, 362, 363
  total body, measurement of 12, 13, 318
  urine excretion *(see kaliuresis)*
Potassium depletion *(see also hypokalemia)*
  acid-base disturbances *(see acid-base disturbances)*
  adaptation to 258, 259
  angiotensin II *(see renin-angiotensin-aldosterone axis)*
  Bartter's syndrome 245, 246, 304
  blood pressure 191, 193-196
  causes
    extrarenal 146, 205-220, 231-238, 261
    renal 205, 239-248
  diarrhea 146, 147, 231-238, 315-319, 397
  diuretics *(see diuretics)*
  hypokalemia *(see hypokalemia)*
  indomethacin, effects on potassium excretion 304, 348
  kaliuresis *(see kaliuresis)*
  magnesium *(see magnesium deficiency)*
  nephrogenic diabetes insipidus 202
  peripheral vascular resistance 191, 193-196
  plasma renin activity *(see renin-angiotensin-aldosterone axis)*
  relationship with plasma potassium concentration 225
  renal adaptation 258, 259

Potassium depletion *(continued)*
  risk of severe manifestations of 278-280
  skeletal muscle 191, 197-201
  "tea and toast" diet 226, 227
  total body potassium 12, 13, 318
  vomiting 232, 233, 325, 326 *(see also hyperkalemia, hypokalemia and kaliuresis)*
  aldosterone effects *(see aldosterone and aldosteronism)*
  and acid-base balance 178-182, 184-186, 202, 265, 300-303, 309-312, 364, 371, 372, 403, 432-434
  and diuretics *(see diuretics)*
  and mineralocorticosteroids *(see aldosterone and aldosteronism)*
  and potassium intake 148, 153, 163-165, 183, 196
  and potassium-sparing diuretics *(see diuretics)*
  and sodium
    excretion 173, 174, 177, 403
    intake 173, 174, 177, 403
  carbonic anhydrase inhibition and 161, 162, 240, 265, 305
  effects of alkali therapy on 124-136, 410, 411
  fractional excretion 182-186
  gastrointestinal 146, 147, 231-238, 315-319, 397
  homeostasis of, and ammonia production 202, 228, 279, 280, 321-325
  in collecting tubule 150-186, 309-312, 371, 372, 403
  in distal nephron 150-186, 309-312, 371, 372, 403
  in renal failure 182, 184-186, 364
  intestinal 146, 147, 231-238, 315-319, 397
  losses of, from skin 146, 261
  plasma potassium levels, effects on 153, 163-166
  regulation of 148, 187-189, 364-389
  renal *(see also kaliuresis)* 146, 148-165, 167-186, 239-248
  urine flow rate 161, 177, 273-275
  urine sodium/potassium ratio 168, 171, 173, 174

Potassium 235

Potassium-hydrogen exchange processes 259
Potassium ions (see *potassium*)
Potassium loading, and kaliuresis 182-186
Potassium metabolism (see also *potassium*)
  acid-base balance (see *acid-base disturbances*)
  adaptation to potassium loading
    extrarenal 165, 166, 397
    renal 165, 182-186, 388, 389
  alcoholism 228, 321-324, 426-429
  aldosterone (see *aldosterone and aldosteronism*)
  alpha-adrenergic agonists and antagonists (see *catecholamines*)
  antidiuretic hormone (ADH) 172, 175-177, 191, 202
  barium (Ba^{++}) effects 90, 91, 212, 219, 306
  beta-adrenergic agonists and antagonists (see *catecholamines*)
  bicarbonate 124-136, 410, 411
  blood pressure 191, 193-196
  carbonic anhydrase inhibitors 161, 162, 240, 265, 305
  catecholamines (see *catecholamines*)
  cellular entry (see also Na^+, K^+-ATPase) 76
  chloride (see *chloride*)
  corticosteroid hormones (see *corticosteroids*)
  dietary potassium intake 392-396
  digitalis (see *digitalis*)
  distribution among body fluids 3, 4, 7, 9, 11, 13-16, 26-31, 34, 43, 57, 108
  electrochemical gradient (see *potential and potential difference*)
  excretion
    feces 146, 147, 231-238, 315, 319
    urine 148-189, 309-312, 371, 372, 403
  food content 393-396
  fractional excretion of potassium 182-186
  glucagon 117, 344
  glucocorticosteroids (see *corticosteroids*)
  glycogen metabolism 1, 203

Potassium metabolism (*continued*)
  hepatic uptake 13, 145
  hypertension, prevention of 191, 193-196
  insulin (see *insulin*)
  intracellular concentration 7, 72
  K^+/N ratio (potassium/nitrogen ratio) 1, 169, 170, 222, 223
  liver 13, 145
  luminal chloride concentration in the distal nephron 152, 155, 162, 173, 174, 177, 239, 240
  luminal sodium concentration in the distal nephron 162, 173, 174, 177, 400
  mannitol 142, 243, 353, 412
  mineralocorticosteroids (see *aldosterone and aldosteronism*)
  muscle (see *skeletal muscle*)
  natriuresis 162, 173, 174, 400
  potassium-sparing diuretics (see *diuretics*)
  potassium supplementation (see *potassium repletion*)
  renal adaptation to K^+ depletion 258
  rhabdomyolysis 347, 359, 426-429, 435-438
  selective potassium channels (K^+Ch) (see *channels, potassium*)
  skeletal muscle (see *skeletal muscle*)
  sodium chloride balance 400
  sodium-induced kaliuresis 162, 173, 174, 177, 400, 403
  sodium-potassium-ATPase (see *Na^+, K^+-ATPase*)
  splanchnic uptake 13, 145
  stores 12, 13
  stress hypokalemia 209, 253, 255, 257
  succinylcholine 347, 351, 352
  total body 12, 13, 318
  transport (see *channels, potassium, and Na^+, K^+-ATPase*)
  water balance 161, 172, 175-177, 191, 202, 273-275, 364-366
Potassium repletion (see also *hyperkalemia, and hypokalemia*)
  beta-adrenergic blockers (see *catecholamines*)

Potassium repletion *(continued)*
 complications *(see hyperkalemia)*
 intravenous administration of potassium 145, 268-277, 283-287, 300-303, 363
 Kaon-Cl 282
 liquid oral forms 282
 micro-K 282
 nitrogen-sparing effect 169, 170, 222, 223
 oral administration of potassium 145, 268-277, 281-285, 287-295, 362
 potassium-sparing agents *(see also diuretics)* 293-299
 risks *(see also hyperkalemia)* 276, 277
Potassium transport *(see also potassium metabolism)*
 acid-base balance *(see acid-base disturbances)*
 active transport *(see ATPase)*
 aldosterone *(see aldosterone and aldosteronism)*
 barium (Ba^{++}) 90, 91, 212, 219, 306
 channels 75, 78-80, 84-94
 cortical collecting tubule 150-186, 265, 311, 312, 371, 372, 403
 distal tubule 150-186, 265, 309-312, 371, 372, 403
 diuretics *(see diuretics)*
 electrical potential difference (PD) *(see potential and potential difference)*
 entry to cells *(see also ATPase)* 76
 exercise 121, 123
 exit from cells *(see also channels, potassium)* 75
 fractional excretion of K^+ in the urine 182-186
 glucocorticosteroids *(see corticosteroids, glucocorticosteroids)*
 in renal tubular acidosis (RTA) 265, 309-312, 317, 372, 403
 mineralocorticosteroids *(see corticosteroids)*
 Na^+, K^+-ATPase *(see Na^+, K^+-ATPase)*
 Nernst equation 39-43, 47, 48, 51
 permeability (P_{K^+}) 64, 63, 71, 77
 principal cell of distal nephron 150, 388
 recycling in the kidney 172, 176

Potassium transport *(continued)*
 secretion
  colon 146, 147, 231-238
  distal nephron 150-186, 265, 311, 312, 371, 372, 403
 sympathetic activation and blockade *(see catecholamines)*
 urine flow rate 172, 175-177, 191, 202, 403
 vasopressin *(see ADH)*
Potassium wasting *(see kaliuresis, hypokalemia, and potassium excretion)*
Potential
 action 70, 97
 diffusion 67, 68
 electrochemical 31-38, 62
 equilibrium 37-54, 85
 intracellular 31, 59, 74, 114
 luminal, in distal nephron 155-160
 resting 1, 49-51, 59, 66, 68-71, 74, 98-100, 104
 threshold 96, 100, 101, 104
Potential difference *(see also potential)*
 cell membrane 31-36, 38-46, 48-53, 66, 96-99
 transepithelial 154-160
Prednisone *(see corticosteroids)*
Propranolol *(see catecholamines)*
Prostaglandins 304, 348, 371, 379, 382, 391
Proteins
 cell-restricted anions 16-18, 21
 Gibbs-Donnan rule 16, 17
 intracellular potassium binding capacity 16, 17
"Pseudo K^+-depletion" 169, 170, 222, 223
Pseudohypoaldosteronism (tubular resistance to mineralocorticosteroids) 384, 386
Pseudo or spurious hyperkalemia 328
Pseudo or spurious hypokalemia 250, 251
Pump-leak hypothesis 15, 30
Red cell
 electrolyte composition 8, 72, 73
 hemolysis and hyperkalemia 72, 344, 347, 353, 359, 423
 potassium content of 13, 72

Renal failure *(see also uremic acidosis)*
  hyperkalemia in 364-389, 397-400, 419-422, 430-434
  hypocalcemia 101, 103, 104, 331, 332
  tissue excitability 101, 103, 104, 331, 332
Renal transplantation and potassium disorders 348, 372, 379-381
Renal transport of potassium *(see potassium transport)*
Renal tubular acidosis (RTA) and hypokalemia
  distal or type-1 RTA 311, 312
  proximal or type-2 RTA 311, 312
  proximal vs. distal RTA 265, 311
Renal tubular acidosis type-4 (RTA) and hyperkalemia 371, 372, 403
Renin *(see renin-angiotensin-aldosterone axis)*
Renin-angiotensin-aldosterone axis
  blood pressure 255, 267, 325, 326
  diabetes 371, 372, 403
  diuretics 244, 267, 293-299, 304
  hyperaldosteronism *(see aldosteronism)*
  hyperkalemia 364-379, 403
  hypoaldosteronism 296, 297, 364, 368-379, 403
  hypokalemia *(see aldosteronism)*
  hyporeninemic hypoaldosteronism 364, 369, 371, 372, 375, 376, 403, 404
  magnesium levels 245, 247, 248, 266, 267
  plasma potassium *(see hyperkalemia and hypokalemia)*
  urinary potassium 171, 400, 403
  urinary sodium/potassium ratio 171
Renin-secreting tumor 325, 326
Respiratory acidosis and plasma potassium 135, 136
Respiratory alkalosis and plasma potassium 132, 135, 136, 211
Respiratory muscles paresis of, in hyperkalemia and hypokalemia *(see paralysis)*
Resting membrane potential *(see potential, potential difference)*
Rhabdomyolysis *(see also hypokalemia, necrosis, potassium depletion)*
  hyperkalemia 423, 426-431, 435-438
  hypokalemia 197-201, 435, 436

RTA *(see renal tubular acidosis)*
Salt in the diet, and kaliuresis 400, 402, 403
Sea, electrolyte composition 3, 4, 6
Serum potassium *(see potassium)*
Shock, and lactic acidosis 137-140, 426-429, 435-438
Sjögren's syndrome and renal tubular acidosis (RTA) 309-312
Skeletal muscle composition
  in normal state 7, 8, 12, 13, 27, 72, 73
  in potassium depletion 198-201
Skin, potassium losses in sweat 146, 261
Sodium
  effect on potassium excretion 162, 173, 174, 177, 365, 366, 400, 403
  extracellular 9, 22, 23, 28, 29, 47-49, 51
  intracellular 8, 22, 47-51, 53, 54, 60
  renin-angiotensin-aldosterone axis *(see renin-angiotensin-aldosterone axis)*
  retention of, in hypokalemia 198, 202
  transport of, and potassium transport *(see Na^+, K^+-ATPase)*
Sodium bicarbonate *(see also bicarbonate)*
  excretion and kaliuresis 152-155, 178-181, 239-241
  therapy 124-136, 410, 411
Sodium chloride *(see sodium)*
Sodium-hydrogen exchange 59, 113, 115
Sodium, potassium-ATPase *(see Na^+, K^+-ATPase)*
Spironolactone 293-297
Spurious or pseudohyperkalemia 328
Spurious or pseudohypokalemia 25, 250
Stool, potassium losses in
  diarrhea 146, 147
  normal state 146, 147
  renal failure 397
Stress hypokalemia 209, 253, 255, 257
Succinylcholine and hyperkalemia 347, 351, 352
Sweat, potassium losses 146, 261
"Tea and toast" diet 226, 227
Tetraethylammonium (TEA) 88
Tetrodotoxin (TTX) 88
Thiazide diuretics *(see diuretics)*

Tissue analysis *(see also red cell, skeletal muscle)* 72, 73
Tissue necrosis, and hyperkalemia 423, 426-429, 430, 431, 435-438
Transplantation, renal
 and cyclosporine 348, 379-381
 and hyperkalemia 372, 384
Transtubular electric potential difference *(see potential and potential difference)*
Triamterene 293-299
Tubular resistance to mineralocorticosteroids (pseudohypoaldosteronism) 384, 386
Tubule
 collecting, potassium transport in 150-186, 265, 309-312, 371, 372, 403
 distal nephron, potassium transport in 150-186, 265, 309-312, 371, 372, 403
Tumors
 ACTH-producing 325, 326
 renin-secreting 325, 326
 villous adenoma 231, 238

Uremia *(see renal failure)*
Uremic acidosis *(see acid-base disturbances)*
Ureteral anastomosis to intestinal tract 236, 237
Urinary electrolytes *(see chloride, kaliuresis, and sodium)*
Urinary tract obstruction *(see renal failure)*
Urine concentrating defect in hypokalemia 202
Urine flow rate, and potassium transport *(see also kaliuresis)* 172, 175-177, 191, 202
Vascular tone, effect of potassium depletion on 191, 193-196
Vasopressin *(see antidiuretic hormone)*
Ventricular arrhythmias *(see cardiac arrhythmias)*
Villous adenoma and hypokalemia 231, 238
Vomiting, protracted, and potassium loss 232, 233, 325, 326
Xanthines and $[K^+]_p$ 209

Order Form

Mail to L & G Publications, Inc.
10703 Paulwood
Houston, Texas 77071

Quantity	Book Title	Price	Total
	Acid-Base	$ 29.95	
	Potassium	29.95	
	Salt & Water *(Sept 92)*	29.95	
	Heart Failure *(March 93)*	29.95	
	Renal Failure *(Sept 93)*	29.95	

Tax of 7.25% applies to Texas residents only.	Subtotal	
UPS ground shipping: $3 for first item, $1 for each additional.	Tax	
UPS 2nd day air: $6 first item, $2 each additional.	Shipping	
Air mail to Canada: $5 first item, $3 each additional. Air mail overseas: $14 each item.	**Total**	

_____ Payment enclosed (check or money order, no cash please)

Name	
Address	
City	
State	Zip Code

Libra & Gemini Publications, Inc.
10703 Paulwood ❖ Houston, TX 77071
713/981-6321

Order Form

Mail to L & G Publications, Inc.
10703 Paulwood
Houston, Texas 77071

Quantity	Book Title	Price	Total
	Acid-Base	$ 29.95	
	Potassium	29.95	
	Salt & Water *(Sept 92)*	29.95	
	Heart Failure *(March 93)*	29.95	
	Renal Failure *(Sept 93)*	29.95	

Tax of 7.25% applies to Texas residents only.
UPS ground shipping: $3 for first item, $1 for each additional.
UPS 2nd day air: $6 first item, $2 each additional.
Air mail to Canada: $5 first item, $3 each additional.
Air mail overseas: $14 each item.

Subtotal	
Tax	
Shipping	
Total	

_____   Payment enclosed (check or money order, no cash please)

Name	
Address	
City	
State	Zip Code

Libra & Gemini Publications, Inc.
10703 Paulwood ❖ Houston, TX 77071
713/981-6321

Order Form

Mail to L & G Publications, Inc.
10703 Paulwood
Houston, Texas 77071

Quantity	Book Title	Price	Total
	Acid-Base	$ 29.95	
	Potassium	29.95	
	Salt & Water *(Sept 92)*	29.95	
	Heart Failure *(March 93)*	29.95	
	Renal Failure *(Sept 93)*	29.95	

Tax of 7.25% applies to Texas residents only.
UPS ground shipping: $3 for first item, $1 for each additional.
UPS 2nd day air: $6 first item, $2 each additional.
Air mail to Canada: $5 first item, $3 each additional.
Air mail overseas: $14 each item.

Subtotal	
Tax	
Shipping	
Total	

_____ Payment enclosed (check or money order, no cash please)

Name	
Address	
City	
State	Zip Code

**Libra & Gemini Publications, Inc.
10703 Paulwood ❖ Houston, TX 77071
713/981-6321**

Order Form

Mail to L & G Publications, Inc.
10703 Paulwood
Houston, Texas 77071

Quantity	Book Title	Price	Total
	Acid-Base	$ 29.95	
	Potassium	29.95	
	Salt & Water *(Sept 92)*	29.95	
	Heart Failure *(March 93)*	29.95	
	Renal Failure *(Sept 93)*	29.95	

	Subtotal	
Tax of 7.25% applies to Texas residents only.	Tax	
UPS ground shipping: $3 for first item, $1 for each additional.	Shipping	
UPS 2nd day air: $6 first item, $2 each additional. Air mail to Canada: $5 first item, $3 each additional. Air mail overseas: $14 each item.	**Total**	

_____ Payment enclosed (check or money order, no cash please)

Name	
Address	
City	
State	Zip Code

Libra & Gemini Publications, Inc.
10703 Paulwood ❖ Houston, TX 77071
713/981-6321

Order Form

Mail to L & G Publications, Inc.
10703 Paulwood
Houston, Texas 77071

Quantity	Book Title	Price	Total
	Acid-Base	$ 29.95	
	Potassium	29.95	
	Salt & Water *(Sept 92)*	29.95	
	Heart Failure *(March 93)*	29.95	
	Renal Failure *(Sept 93)*	29.95	

Tax of 7.25% applies to Texas residents only.
UPS ground shipping: $3 for first item, $1 for each additional.
UPS 2nd day air: $6 first item, $2 each additional.
Air mail to Canada: $5 first item, $3 each additional.
Air mail overseas: $14 each item.

Subtotal	
Tax	
Shipping	
Total	

_____ Payment enclosed (check or money order, no cash please)

Name	
Address	
City	
State	Zip Code

**Libra & Gemini Publications, Inc.
10703 Paulwood ❖ Houston, TX 77071
713/981-6321**